Design Science Research Methodology

José Osvaldo De Sordi

Design Science Research Methodology

Theory Development from Artifacts

José Osvaldo De Sordi
Business School
Federal University of Sao Paulo
Osasco, São Paulo, Brazil

ISBN 978-3-030-82155-5   ISBN 978-3-030-82156-2   (eBook)
https://doi.org/10.1007/978-3-030-82156-2

The translation was done with the help of artificial intelligence (machine translation by the service DeepL.com). A subsequent human revision was done primarily in terms of content.

Cover illustration: © John Rawsterne/patternhead.com

This Palgrave Macmillan imprint is published by the registered company Springer Nature Switzerland AG
The registered company address is: Gewerbestrasse 11, 6330 Cham, Switzerland

*To my wife Keller and
my daughters Isabel and Carolina.*

FOREWORD

At the dawn of the twenty-first century, dominated more and more by digital technologies and by a deluge of information, which aim at supporting functions at organizations, we are still faced with the fragility in the development of information systems without a scientific basis. This fragility has immense costs. Not only the cost of failing investments in information systems, but also of failed promises of innovative changes in the organization, often with huge impact on professionals' motivation and trust.

The Design Science Research Methodology (DSRM) is a relatively recent approach, an actual result that emerged from the 2000s scientific struggle to address the problem of designing and implementing information systems in a scientific manner. It is therefore already a result of this new century.

Design science "foundation" owns to the previous works by Herbert A. Simon (stablished by its masterful book *The Sciences of the Artificial*) and Richard Baskerville and others. Although the concept of "design science" has been introduced much earlier by R. Buckminster Fuller in 1957.

It is with the paper about "Design Science in Information Systems Research" in 2004, that Alan Hevner and colleagues brought the attention of this approach to the researchers and practitioner of the information systems arena. With this work an approach to use design science in the definition of a set of artifacts related with information systems, from models to actual instantiations was established.

Moreover, it was "formally" acknowledged the necessity to use science in designing and developing information systems, leading to the recognition of the complexity involved in constructing an information system. At around the same time, Kenneth and Jane Laudon, with their anthological book on *Management Information Systems*, further explains that an information system should be about dealing with people, (working) processes and (supporting) technology. So, from this time on, researchers and practitioners had in their hand a more practical, and I should add pragmatic, "tool" to deal with real-world (scientific) information systems problems. With this "pragmatist" epistemological perspective in our hands, another door opens to research. Contrasting with the positivistic research concept, whereas one states an objective knowledge acquired by examining empirical evidences and hypothesis testing, and different from the constructivists, where one proposes that knowledge is relative and that reality is too complex to be understood, the "pragmatic" approach believes that the process of acquiring knowledge, like in the real world, is a continuum (e.g., often a recurrent cycle). Therefore, pragmatism can be situated somewhere in the center of the paradigm continuum in terms of mode of inquiry, that leads to an epistemology that enables an "engineering or design way" to improve the world. More than looking for the "truth," DSRM is aiming at improving the world with a systematic process to design "artifacts" purposely built to solve specific "real-world" problems.

One of the most interesting areas of recent application of DSRM with significant success is the complex field of health information systems, where the paper by Lapão and colleagues (Implementing an online pharmaceutical service using design science research), in 2017 became among the first examples, as recognized by the recent Hevner paper on "Design Science Research Opportunities in Health Care" (2018). DSRM is conquering more and more fields with its pragmatic research paradigm, perhaps better fitted to a more and more complex world.

However, one of the most noteworthy aspects of applying DSRM in the healthcare is to leverage its design features to promote a participatory approach to the design and implementation of health information systems. More and more, and motivated by an "epidemics" of information systems implementations failures, designers are looking at extending further the DSRM to include and engage the stakeholders in the design process. Precisely like what most architecture school recommend by appealing to the contributions by the future building owners. Simões and colleagues (2018), applied DSRM to support the participatory implementation of an antibiotic stewardship information system, clearly showing the benefits of early engaging the stakeholders to promote a better translational process and permitting a wider professional's ownership.

The scientific translation for supporting the implementation of information systems is a key aspect where DSRM is playing a positive role, as one can look at DSRM as a valid approach to sustain new information systems, embracing the DSRM cycle, from design to implementation and even sustainability phases.

The present Book, by Professor De Sordi, also enlightens the way DSRM interacts with the context where it is used, with clear benefits for the translation of knowledge into the real world. The book starts with an engaging and comprehensive description of the DSRM theories, epistemological context and history of the establishment of the field of design science.

It is followed by a rich and broad list of DSRM papers' publications and the journals where they were published. This list is a complementary base of information to a "real-world" connection with the history and recent developments of DSRM. One of the most valuable things about this book is how it addresses the practice of DSRM and its challenges, with a set of illustrative and pedagogical examples. These examples can be very useful to further understand the benefits of adopting DSRM to tackle the development of information systems.

For the information systems researchers, the chapter about the impact of the research is almost obligatory to motivate the adoption of DSRM and to involve more colleagues and students in this field of increasing importance in the field.

The chapter on Artifacts development is central in the book, emphasizing the extended role of the artifacts in design science. From real-world instantiation (Level 1) to more general theories (Level 3), DSRM contributes to create knowledge while improving the world with new proposed solutions. However, it is pointed out that the existing gap, identified by Baskerville and Colleagues, between Level 1 and Level 3 research is indeed an opportunity to be explored by DSRM researchers. Nevertheless, and very interestingly from Professor De Sordi own research, it seems that the skills needed to develop an artifact (Level 1), are not the same as those required to develop a theory (Level 3). This opens a window of an opportunity to bridge the gap between level 1 and 3 contributions to DSRM.

This chapter is followed by the one that further explains the increasing use of DSRM, even beyond the standard approaches to information systems, leading to complete the analyse elucidating a set of recent variations of DSRM, focusing mainly on the Action Design Research and Ground Design.

The book ends with a very interesting examination of the communication dimension of DSRM. Communication is an often forgotten essential component of DSRM, critical for providing researchers and practitioners to share the knowledge obtained in the construction of the artifact. This chapter further adds some hints on how to leverage this component. Furthermore, the book also delivers some annexes, including one with a practical checklist on how to develop a DSRM approach. From my side it was pleasure to read through it and I wish the reader a similar experience, and I am sure the reader will come back more times to get relevant hints for his or hers research on DSRM.

May 2021

Professor Dr. Luís Velez Lapão
Professor of Global Health Information
Universidade Nova de Lisboa
Lisbon, Portugal

Acknowledgements The development of the sixth chapter, Theory Development from Artifacts, was only possible thanks to the help of my colleagues and researchers Marcia Carvalho de Azevedo (Federal University of São Paulo), Luis Herman Contreras Pinochet (Federal University of São Paulo), and Carlos Francisco Bitencourt Jorge (University of Marília), who assisted me in the collection and analyses of the many articles that applied the design science research method. To them my gratitude and affection for the great dedication of time and analytical efforts.

CONTENTS

LIST OF FIGURES

LIST OF TABLES

CHAPTER 1

DSR as an Innovation Accelerator Strategy

CHALLENGE OF THE TRANSITION FROM INVENTION TO INNOVATION

The concept of invention is usually associated with an original idea, which theoretically makes sense and works. The word theory refers to science and to the typical competences of the researcher, encompassing knowledge, skills, and attitudes employed by the scientific community. Inventions usually take place in research centers and in research and development (R&D) areas, traditional locus of researchers' activities. Innovation is associated with the implementation of this original idea, so as to turn it into a product or service that delivers value to a community of people (customers). At this point, other areas and skills are required such as marketing, costs and pricing, product/service engineering, strategy, among others. The long time interval from invention to innovation is usually a challenge, a problem intensely discussed by academia and with many denominations, among which the "problem of translation" (David et al., 1992).

Haeussler and Assmus (2021) studied the "problem of translation" in the context of the pharmaceutical industry, having drugs for sick people as a result of innovation. They observed that the results of the transition from invention to innovation are better when the responsible researcher has the "ability to translate," composed by the domain of vertical experience and horizontal skills. Vertical experience is related to knowledge

© The Author(s), under exclusive license to Springer Nature Switzerland AG 2021
J. O. De Sordi, *Design Science Research Methodology*,
https://doi.org/10.1007/978-3-030-82156-2_1

obtained from the study of several diseases, while the horizontal ability of the researcher is in having the ability to develop both basic research and applied research. Horizontal skills are considered by the authors to be the main factor in solving the problem of translation. Thus, the ability of researchers to carry out both types of research, while also developing applied research, is fundamental not only to reduce the time from invention to innovation, but also for the transition to occur, that is, to solve the "problem of translation."

The difficulties in overcoming the "problem of translation" are many, configuring itself as a complex problem that has persisted for decades. One way to portray this difficulty is by presenting the differences between texts aimed at the scientific public and those aimed at practitioners. Texts for practitioners are those directed to professionals in areas of science known as professional schools, such as medicine, administration, law, engineering, among others. Next, we describe the symmetrically opposed facilities and difficulties between the scientific literature and the literature directed to practitioners.

Scientific Literature

Scientific texts in general are not easy to be read and understood by an audience other than researchers. Even the scientific texts of the so-called professional schools, such as medicine, law, engineering, administration, among others, are not likely to be read and directly applied, that is, readily "consumed" by practitioners of the area. They are texts aimed at constructions or criticism of theories. These texts explore, describe and/or explain generally new themes. The attention is on the discussion of new and scientifically valid knowledge, there is no concern in the exercise, in the application of that knowledge by others than the researchers of the area in question. One unknown term often makes the reader have to stop reading to access another document from the citation. In general, many terms require consultation by practitioners, considering the triangulation of theories, strategies and techniques in the scientific text is very common. Another difficulty is that a scientific advance is usually produced slowly and gradually, the result of the evolution of different researches, published in many articles over time (Van Aken & Romme, 2009). Thus, to have the understanding of a scientific knowledge, the practitioner must perform a lot of reading of these researches that are fragmented in different articles.

On the other hand, the positive aspect of scientific texts is the quality and safety of the information present. The double blind review procedure brings this advantage to the scientific writing, the quality of the data collected, as well as their analyses. The texts present a simple and direct writing, its information is verified by peers, that is, there are no poorly grounded guesses as there are in commercial literature or in texts produced and oriented to practitioners.

Literature for Practitioners

Texts oriented to practitioners, such as books covering techniques and methods for groups of practitioners, are very prescriptive, with lots of guidance on how to do it. Many refer to these as self-help books. The biggest criticism for these texts is that there is little guarantee in terms of the accuracy, precision and veracity of their recommendations. The tests often do not exist or are not exhaustive and do not present the same rigor of those present in scientific texts. They are originated more from the practical experience of their authors, i.e., they do not benefit from the thorough application of the method and the criticisms and advice of reviewers, as occurs with the scientific text (Van Aken & Romme, 2009).

Scientific texts are rich in evidence but poor in solutions; in the opposite situation, texts oriented to practitioners are rich in prescribing solutions on how to do it, but with little evidence of rigorous data collection and analysis,, reflecting in texts of dubious quality. Figure 1.1 depicts this dilemma, this mismatch between these two types of literature. In a way, it helps to better understand part of the difficulty of the transposition of the scientific invention to the context of business innovation.

Contributions of the Pragmatist Research Paradigm

According to Creswell (2014) are four types of paradigms for generating scientific knowledge: postpositivism, constructivism, transformative, and pragmatism. The pragmatism paradigm is described as being real-world practice-oriented, pluralistic, consequences of actions, and problem-centered. These characteristics bring the actions of researchers and their scientific discoveries closer to what happens in the work environment of practitioners, whether in medical offices, mechanical workshops, courts of law, or any other workplaces of professionals trained by professional

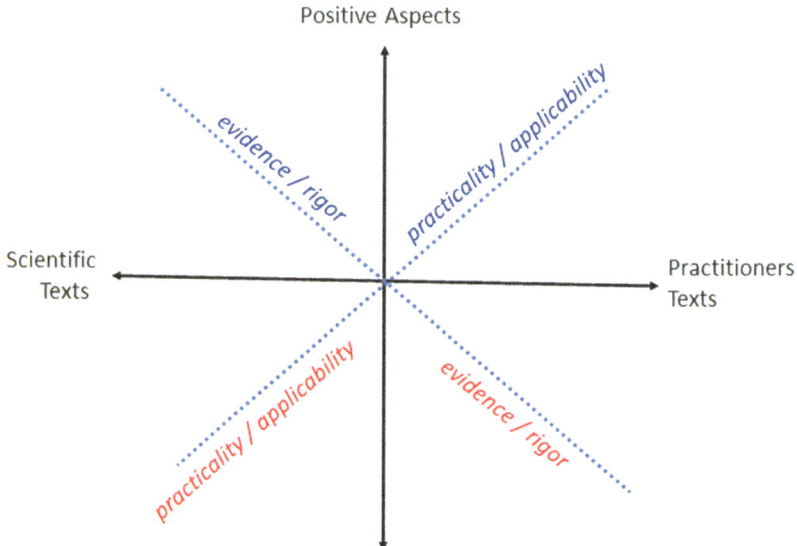

Fig. 1.1 Positive and negative aspects of the scientific and practitioner literatures

schools. Being based on a sensitive problem, linked to the reality of the practitioner, the scientific texts developed with the pragmatic research strategies are more intelligible and useful for the practitioners, as well as assist in reducing the time gap between invention and innovation. Here we have an interesting aspect to be observed, these are researches that are developed already within the environment, the context of real-life application.

Two research strategies within the pragmatic research paradigm are: the Case Study and Design Science Research (DSR), the latter the central object of discussion of this book. The greater proximity and relevance of research developed with these strategies to the interests of practitioners is highlighted even in the way of disseminating their results. In the case study strategy it is recommended that researchers are not restricted to the generation and publication of scientific articles, but that they also disseminate their findings to other stakeholders, including practitioners. Yin (2014, p. 268) recommends the preparation of "a popularized version - appearing in a practitioner journal" that is fun and easy to read, exactly the opposite of the scientific article. In the DSR strategy, it is recommended that the research results are effectively exposed to the technological and business audiences, relevant to the artifact (Hevner et al., 2004).

Briefly, we will address a research strategy in this book that can collaborate with the reduction of time from invention to innovation, by collaboratively bringing the researcher closer to the public that has the demand, i.e., the one that feels the real world problem. Since it is problem-focused, it is obviously valid in terms of being useful, of meeting something that is really necessary. The idea here is to bring together the positive aspects of both types of literature: the practicality and prescription of the practitioner literature, with the rigour and validity of the scientific literature.

References

Creswell, J. W. (2014). *Research design: Qualitative, quantitative, and mixed methods approaches* (4th ed.). Sage.

David, P. A., Mowery, D., & Steinmueller, W. E. (1992). Analysing the economic payoffs from basic research. *Economics of Innovation and New Technology, 2*(1), 73–90.

Haeussler, C., & Assmus, A. (2021). Bridging the gap between invention and innovation: Increasing success rates in publicly and industry-funded clinical trials. *Research Policy, 50*(2), 104155.

Hevner, A. R., March, S. T., Park, J., & Ram, S. (2004). Design science in information systems research. *MIS Quarterly, 28*, 75–105.

van Aken, J. E., & Romme, G. (2009). Reinventing the future: Adding design science to the repertoire of organization and management studies. *Organization Management Journal, 6*(1), 2–12.

Yin, R. K. (2014). *Case study research: Design and methods* (5th ed.). Sage.

CHAPTER 2

Core Concepts of DSR

SCIENCE OF THE ARTIFICIAL

A first step in understanding the DSR is to understand the distinction between natural science and artificial science. Natural science focuses on the study of physical and abstract entities of the world, exploring their characteristics, properties, behaviors, and interactions among them. The artificial science is dedicated to the study of entities developed by man with the purpose of providing some functionality in order to solve limitations or problems. These human-developed entities are generically called artifacts (Simon, 1996). Artifacts are part of the very definition of human being, as highlighted by the historian and essayist Thomas Carlyle: "Man is a tool-using animal. Without tools he is nothing, with tools he is all." Since the first artifacts, such as the cutting instrument made of stone, the mastery of fire, our life is guided by the daily use of artifacts. Thus, the artificial things have a fundamental role in our lives and, therefore, totally pertinent to science to have them as an object of study.

An important aspect of the artificial science is the idea of the contextualization of the artifact, its environment of use, characterizing who are its end-users or practitioners who will use it. The beneficiaries of the application of the artifact are not always its operators or practitioners, these can be intermediaries, and they operate the artifact to generate the benefit to a third party. A typical example is the physician who makes use of several instruments during surgery, having as main beneficiary the patient. In this

© The Author(s), under exclusive license to Springer Nature
Switzerland AG 2021
J. O. De Sordi, *Design Science Research Methodology*,
https://doi.org/10.1007/978-3-030-82156-2_2

case both doctor and patient are in the context of the action, but with different roles: practitioner and beneficiary, respectively. There may be several forms of involvement of several other entities outside the context of the action, for example, the class association of physicians indicating restrictions or recommendations on how to perform a certain procedure, the public agency responsible for health in a given country indicating the types of surgeries authorized and those not authorized. Briefly, we have the context of use and application of the artifact and we have the context of the external environment, important for those who propose or study the artifacts. Newell and Simon (1972) portrayed well these two contexts when discussing Human problem-solving, calling the internal space of action inner box and the external space outer box.

The design science research (DSR) approach can be practiced by researchers from all areas of science, including those associated with hard science, considering that they use and depend on artifacts to perform their activities. In this book we will focus on the use of DSR by researchers who work in areas of science linked to professional schools, such as medicine, business, engineering, social work, among others. This, due to the greater applicability and relevance of the artifacts, not only for the researchers' own use, but mainly by providing tools for the professionals who will use them. As highlighted by Van De Ven (2007, p. 1) the "central mission of scholars in professional schools is to conduct research that both advances a scientific discipline and enlightens practice in a professional domain (Simon, 1976)." Thus, the DSR has great potential to cooperate with professional schools because it is an approach that combines theory with practice.

Focusing on the artifacts of interest of professionals who work at professional schools raises the issue of the specificity of the public customers to be served by these professionals: patients served by doctors; children to be literate by educators; people who want equipment with better performance and performance of engineers; among several other collective demands served by specific professionals from professional schools. Researchers associated with professional schools add more value to society when they act collaboratively with professionals in the area in which they research (Van De Ven, 2007). This leads to a better understanding of the problems in the field, resulting in more appropriate and effective artifacts. In this sense, Simon (1996) commented on the difficulty of traditional sciences in supporting professional schools in terms of providing direct support to the work of their various professionals.

While the traditional sciences focus on the truth, Simon highlights that the design theory, which underlies the DSR approach, has as its main objective the utility (Hevner et al., 2004).

DESIGN THEORY

In terms of philosophy of science, the DSR applies a new concept in terms of the researcher's intentions. The focus shifts from the necessary truth of traditional science to contingent truth. While necessary truth must be shown to be true in all localities and contexts, contingent truth is true in the way it happens or the way things are, but it need not be an absolute and broad truth in all localities and contexts. Thus, we have that the direct outcome theoretical contributions of DSR are termed "mid-range theory, whose validity is limited to a certain application domain" (Van Aken & Romme, 2009, p. 8). As we will observe later, artifacts need to be valid and useful for a specific context, the one characterized by the problem space and the group of professionals (practitioners) outlined by the researcher. According to Simon (1996, p. xii) scientific knowledge, applied contingentially, is summarized in the concern with the design of the solution: professional schools "are concerned not with the necessary but with the contingent not how things are but how they might be in short, with design."

The term design implies in designing something as a solution to a need, and these human creations are called artifacts. The function of the design according to Simon (1996, p. 114) is "devising artifacts to attain goals." A design or the proposition of an artifact follows the same logic of a predictive hypothesis: the designed artifact is our premise (*case*) that we understand to be able to meet our needs (*result*) according to our theoretical understanding (*rule*). It is the typical application of abductive logical reasoning, where from the terms "rule" and "result" a "case" is imagined that can solve the problem at hand. We can make an analogy with the generation of hypotheses of the cause–effect type, where we make inferences about a satisfactory explanation for a specific consequence that we aim at (Lee et al., 2011).

According to Van De Ven (2007, 98) "abduction begins by recognizing an anomaly or breakdown in our understanding of the world, and proceeds to create a hypothetical inference that dissolves the anomaly by providing a coherent resolution to the problem." Abductive logical reasoning is the most commonly exercised by researchers seeking innovative and pragmatic solutions to problems, as well as by practitioners of professional schools for known solutions to known problems. Below I exemplify the physician's abductive reasoning, who seeks to identify the most feasible premise (case) to solve a health problem based on test results and his theoretical framework (rule):

- Rules—the doctor has studied and is aware of the typical symptoms of meningitis, such as sudden high fever, severe headache, stiff neck, vomiting, convulsions, nausea, mental confusion and difficulty concentrating;
- Results—before examining the patient the doctor looks at the results of the tests done before the consultation and listens to the mother's accounts of what happened to her child during the last few hours;
- Case—from the findings of the physical examination and the patient's recent history, the doctor works on the assumption (case) that this patient has a high chance of having meningitis. He then decides to recommend the typical treatment for meningitis to the child.

To better understand the DSR method it is important to be aware of its foundation, for this we will explore the design theory. A theoretical explanation that helps in this sense is the "C-K theory" which is a rigorous and unified formal structure focused specifically for the design (Hatchuel & Weil, 2009). C-K theory addresses two concepts: the space of concepts (C) and the space of knowledge (K). These were defined by Hatchuel and Weil (2009, p. 181) as: "Space K contains all established (true) propositions (the available knowledge). Space C contains 'concepts' which are undecidable propositions in K (neither true nor false in K) about partially unknown objects x." As for the concept entity, it is defined as the figure of an object that exists in the human mind or in the real world, which can be described according to its attributes and function (Taura & Nagai, 2012). In DSR, the concepts associated with the artifact, called space C, are characterized by the undecidable proposition, i.e., they are configured as a contingent truth and do not need to be configured as an absolute and ample truth in all locations and contexts. The important thing is that the concepts of space C are valid and useful for the context of the functionality required for that situation, called in the DSR as that space of the problem.

It should be noted that DSR starts from scientific knowledge (space K) intertwining it with concepts (space C) to compose and deliver innovative artifacts to practitioners or organizations that face problems. A question in this regard would be: how many and which concepts to use? The development of the theme concept generation for design creativity by Taura and Nagai (2012) brings some relevant discussions for a better understanding of this questioning. They decompose the concept generation process into two phases: (a) the *problem-driven phase*; and (b) *inner sense-driven* phase. For the first phase, the main ability is the analysis and the base to be explored is the problem. Thus this first phase brings the awareness between the existing situation and the goal to be achieved. In the second phase, the *inner sense-driven phase,* the main ability is the composition and the base to be explored is the inner sense. In the "inner sense-driven phase is assumed to be related to the process of approaching an ideal through the composition of elements" (Taura & Nagai, 2012, p. 14). The elements are here the concepts to compose a set of integrated functions and capable of delivering the desirable effect.

Similarly to the construction of new theories, which is derived from many triangulations of theoretical perspectives on the same data set (Yin, 2014), the development of an artifact in the DSR is the result of the triangulation between different concepts and knowledge. For the generation of concepts in DSR, Taura and Nagai (2012) emphasize the importance of the researcher in identifying the attributes (properties) of each concept so as to better understand its function, thus cooperating with the identification, differentiation and selection of the appropriate concepts to integrate and compose the solution to a given problem. The importance of attributes is similar to that discussed in the various research approaches. In grounded theory, for example, in the open coding phase, the objective is to find concepts and the analysis of attributes is essential to structure concept categories based on their similarities and differences (Corbin & Strauss, 1990).

As a way to illustrate the composition of concepts for the creation of an artifact, Taura and Nagai (2012) gave examples of physical artifacts and people's everyday life, making it more understandable to researchers and professionals from different areas. One of the examples is the design of an art knife conceived from the junction of two concepts: broken glass and chocolate segments. These concepts embodied the design of a cutting blade composed of segments, like the divisions present in a chocolate bar, which can be easily broken, as well as a thin sheet of ice. As a result, the knife always remains with the possibility of multiple sharp cuts, as many segments are available along the length of the blade.

Throughout the chapters of this book I will use as an example one of the artifacts I developed using the DSR approach: extensive texts cohesion analysis approach or AnaCoTEx (acronym of the original name in Portuguese: "Analisador de Coesão para Textos EXtensos") (De Sordi et al., 2016). The artifact is composed by procedures and algorithms coded in software, with the aim of analyzing cohesion between the text sections in extensive documents, pointing to correct the sections of disconnected texts. The artifact analyzes the cohesion of the extensive text through procedures and algorithms developed from the interrelation of two concepts:

(a) Cross-referencing concept Cross-referencing − is a reference within a text to another part of the text (Collins, 2013). A cross-reference should indicate to the reader the location in the text where the topic has already been addressed. A simple and practical way to represent cross-references is by using brackets: a single digit in the form [n.] indicates the chapter n; [n.p.] indicates the subchapter p in chapter n; [n.p.q.] indicates the subsubchapter q from the subchapter p in chapter n; this pattern is followed for other items of the structure in texts. Eco (1989, p. 125) argues that cross-references are intended not only to avoid repetition, but also to show the cohesion of the text: "a well-organized thesis should be full of cross-references";

(b) Concept matrix of relations − central element of the technique of network analysis (NA) that points out the interrelationship between the actors of a network. In the case of the AnaCoTEx artifact, each chapter, subchapter and other levels of the structure are understood as actors of the network that interrelate composing a text matrix. NA is useful to demonstrate the connections and relationships between actors. Mathematically, using graph theory to generate network indicators and diagrams, NA allows for the analysis of a network from the relations between its actors (Wasserman & Faust, 1994).

Another similarity of the DSR with other scientific approaches, especially those of the transformative research paradigm (Creswell, 2014), is the insertion of the researcher in the area of application of the artifact. In this sense, the researcher must have a broad domain of the problems of the area, especially the one that is the object of action of the artifact being proposed, as well as the artifacts already available and in use by professionals (practitioners) of the area. Figure 2.1 presents the demands perceived by me and my research colleagues when detecting the opportunity to propose an artifact focused on the analysis of the cohesion of extensive text. This demand is a very common problem among professionals who deal with the development and evolution (versions) of extensive texts, such as consulting and audit managers, editors, and researchers.

The researcher's need to know the difficulties of the field to which the artifact aims to solve, brings us back to the idea of engaged research, of

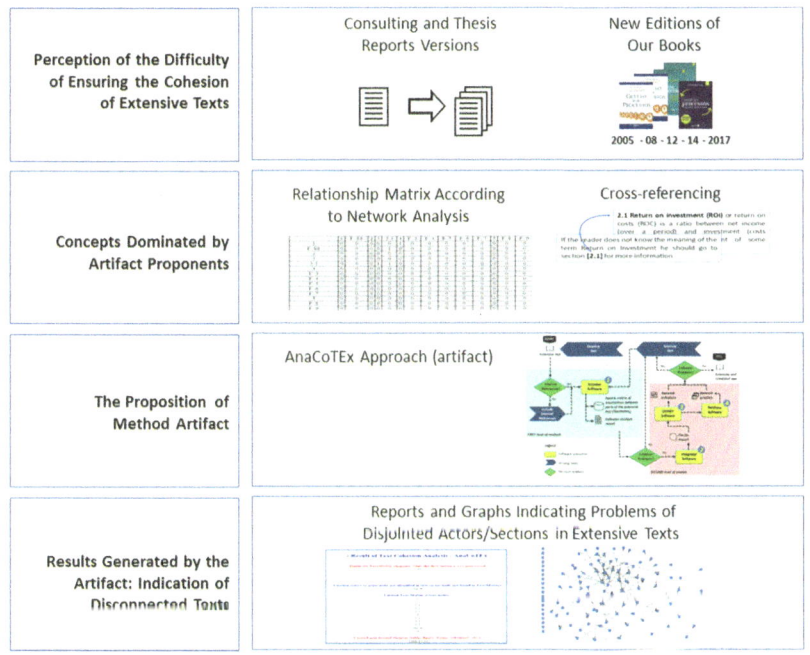

Fig. 2.1 Demand perception, concepts, and idealized artifact

the artifact designed for an audience and a problem of broad domain and knowledge of the researcher. Instead of the distancing of the researcher in relation to the object of study, i.e., the research prepared "for" such demand, here we have the researcher inserted in the context, working "with" the professionals who experience the difficulty. The testing of the artifact in the field, used by typical practitioners, in real field situations (not in laboratories or other environments of researcher control) is the highest expression of this collaborative work. In Chapter 5 we will discuss the issue of using the artifact in the field, in a real-life situation, operated by the practitioners themselves. In the example described in Fig. 2.1, my colleagues and I routinely experienced the difficulty of managing new versions of extensive documents. The demands came from many situations of professional performance, such as the preparation of reports of extensive consultancies, involving multiple phases; editing new books; and reviewing and analyzing the doctoral theses that we supervised.

The proximity of the researcher with the reality of the group of practitioners of the artifact brings several benefits, one of them is to avoid the proposition of unnecessary or useless artifacts. For Hevner et al. (2004) a new artifact does not make sense when: it does not holistically and rigorously meet all the necessary dimensions (financial, ergonomic, and environmental,...); the new artifact does not solve the problem; the existing artifacts are adequate; the utility cannot be proven; or the utility cannot be evidenced clearly and objectively.

TYPE OR CLASS OF PROBLEM

An artifact in DSR can provide a nascent design theory (Gregor & Hevner, 2013) that implies the design of the specificities (meta-requirements and meta-designs) of an artifact directed to a specific situation. By this we mean that it is not something comprehensive as a necessary truth of traditional science, but also not something idiosyncratic to meet the demands of a specific entity. The design of a comprehensive artifact, with multiple functions or "does it all" within the professional schools are not usually adequate. Taking as an example the amphibious cars, they end up having inferior performance in both fields of action, in water and on land. I often compare the artifacts with very comprehensive proposals (multifunctional) and bold with the athletic performance of the duck: flies badly, swims badly and runs badly. Remembering that the DSR artifact must be innovative or outperform existing analogues.

Thus, DSR works within the specificity of artifacts, not so broad and not so specific. As Van Aken and Romme (2009, p. 8) well defined: "it is not a specific solution for a specific situation, but a general solution for a type of problem." We return at this point to the concept of contingent truth, that is, valid and useful artifacts for a specific context, for a type of problem or better describing, characterized and directed to a problem space.

Thinking in terms of class or type of problem equates well with the issue of comprehensiveness, of not being something so macro and not so micro. It is a generalized solution to a specific type of problem. It is not a specific solution for a specific problem, but a conceptual solution whose design specification addresses a type or category of problem, which seeks to address a category of professionals or a business context (not necessarily linked to a specific group of professionals, considering that not every problem or business challenge is specific only to one group of professionals).

Let us take as an example the business challenge of measuring companies' intellectual capital. The problem was manifested by the growing difference, at the end of the twentieth century, between the values declared by companies in financial statements (balance sheet) and their market value. As justification for this difference, the value of the intellectual capital of the companies was identified. Thus, a very specific challenge faced by accountants of large companies was configured. The solution to the problem was by means of artifacts, in the form of methods, that would enable the ascertainment of the value of the companies' intellectual capital. Thus, several methods were developed to face this problem. Jurczak (2008) conducted a review and developed a typology concerning these different methods. He identified 21 methods that were classified into a taxonomic tree with four types: Direct Intellectual Capital Methods (DICM), Market Capitalization Methods (MCM), Return on Assets Methods (ROA) and Scorecard Methods (SC). Thus, we herein have the indication of a type of problem and of a very specific and delineated group of practitioners, respectively, measuring corporate intellectual capital and accountants of large companies.

One aspect to be observed is that the inspiration for an artifact is not always a problem, it can also be an opportunity. Although rarer to occur, an invention can be perceived, even resulting in an artifact previously unnoticed and not demanded by practitioners. Instead of a problem class we might then have an opportunity identified. Artifacts linked to opportunity or innovation are not only rarer to occur, they are also more difficult to publish. As has long been known, in science editors and reviewers are more sensitive and favorable to publishing papers on known problems than solutions for exploring opportunities or avoiding potential problems. Similarly, companies and professionals, who will test artifacts, are also more problem-oriented. According to Ahuja and Lampert (2001) large firms tend to be more resistant to introducing innovations because of: (a) preferring the familiar, over the unknown; (b) preferring solutions closer to existing approaches, over totally new ones. Thus, the most frequent and successful DSR is usually directed toward problem-solving.

Usefulness of the Artifact

As Hevner et al. (2004, p. 98) pointed out "the design-science paradigm seeks to create 'what is effective'." For the researcher who is proposing an artifact with the DSR approach is fundamental reflections about two fundamental questions: (a) "What utility does the new artifact provide?", and (b) "What demonstrates that utility?" An editor or reviewer of a paper developed with the DSR approach will pay attention, among many things, to these two questions. They are structuring questions for research projects designed with the DSR approach.

To answer the first question, What utility does the new artifact provide?" The researcher has to have the clarity of the problem space to be able to clearly spell out the functionality to be delivered by the new artifact. The space race that occurred in the second half of the twentieth century was prodigious in the sense that it easily characterizes the problem space, in the literal sense of the word. The functional challenge of the new artifact, the rocket, was simple enough to be understood by all: to take man from Earth, put him on lunar soil, and bring him back to Earth in good physical and mental condition. Unfortunately, not every definition of an artifact's function originates from a long time collective desire. Usually the researcher has a strong insertion in the field of application of the artifact, i.e., is a practitioner and/or a scholar of the area, thus being a connoisseur of the problem space.

The second question, "What demonstrates that utility?" It is often answered by making use of indicators already available in the application area itself. Let us imagine that we are proposing an artifact in the form of a cutting tool for cranial incisions to be used by neurosurgeons. Let us imagine that one of the neurosurgery modalities, for which the artifact is intended, has a mean post-surgery hospital stay of 10 days and a risk of bacterial infections ranging between 5 and 12%. These two indicators, post-surgery rest time and hospital infection rate, may be two good indicators for analysis and demonstration of the new artifact. Its natural utility is justified not only by the pertinence, but by the wide knowledge and use by the artifact's practitioners.

Unfortunately, DSR artifacts are not always proposed for functions where we can quickly and naturally identify one or more indicators. One situation is when we are proposing innovative artifacts. Returning to the example of the artifact for extensive texts cohesion analysis approach (De Sordi et al., 2016), although the problem was experienced by many practitioners (consultants, auditors, editors, and researchers), there was no indicator in use among practitioners for cohesion analysis between parts of an extensive text. The solution was to create an indicator or reuse some indicator linked to the constructs used for the development of the artifact. The second option was the one that proved to be more feasible. We adopted one of the indicators associated with one of the constructs that made up the artifact, the technique of network analysis. The selected indicator was the network density (d),

$$d = \frac{l}{((n(n-1)/2)} \text{ (Wasserman \& Faust, 1994)} \qquad (2.1)$$

where, n is equal to the number of network actors, in the context of the artifact in question, each of the chapters, subchapters, and other parts of the extensive text; and l is the quantity of existing relationships between the network actors, in the context of the artifact, the invocations between parts of the long text, either formally citing a part of the text or even one of its non-textual elements, such as figures and tables.

For artifacts whose results may also entail undesirable side effects, we should develop indicators to monitor these possible undesirable aspects. This will allow a holistic view of the artifact performance. Thus, the availability of a DSR artifact may be associated with the delivery of several performance indicators that will help to verify the usefulness of the artifact.

EVIDENCE OF USE AND USEFULNESS

The artifact should be field tested in the natural working environment of its typical practitioners. Thus, from the first version of the artifact, used for initial testing, the researcher should have an operational version. As we will see in Chapter 5, "Design Science Research Method," the improvement of the artifact is interactive from feedback from practitioners, from practitioners' clients (beneficiaries of the artifact's action) and from the data generated (outputs) by the use of the artifact itself. The important to be highlighted here is the record, the log of the data resulting from the action of the artifact and coming from one or more sources. With the Internet of Things (IoT) the data collection of these various entities or processed resources has become simpler and more direct. The data records must characterize the conditions of the entity/resource in at least two moments: before the use of the artifact and after its use. With this we can discuss whether the artifact promotes the transition from status A to status B, i.e., if it is able to travel the distance from the current situation to the desired situation that characterizes the problem space for which the artifact was designed.

For the DSR project already mentioned and which I am using as example, the extensive texts cohesion analysis approach (De Sordi et al., 2016), we collected data from the network density of 41 reports (book, Consulting report, research report, thesis, dissertation) before the actions identified and recommended by the artifact and after the changes made by the writer of the extensive text. In the case of this artifact, the chapters and subchapters of the reports were treated by the artifact (software) as a sequence of decomposed numbers. For example, if chapter five has three subchapters, the software will identify four actors: [5], [5.1], [5.2] e [5.3]. Here is a situation that there was no risk of data exposure, considering that we dealt for each of the 41 processed instances (reports) with two text matrices, the initial and the posterior. The two square matrices composed only by numbers, without risk of inference that could expose any content of the analyzed reports. The greatest concern remained in maintaining the anonymity of the names of the authors of the long texts, as well as the titles of the reports administered by them.

As we are talking about the use of the artifact not by the researcher in his laboratory, but by practitioners in their natural working environments, we are dealing with sensitive data, of entities/resources that actually exist. Thus, these data deserve all the attention and care in terms of ethical and responsible treatment. As the advancement of technologies for data collection and storage has evolved greatly in recent decades, due to various technological movements (*e-services*, *pervasive computing*, Industry 4.0, and IoT), concerns about the ethical treatment of data are growing, as well as the solutions, the artifacts made available for such.

For the researcher who adopts DSR with large-scale testing, the care with the data from field testing should be thorough, similar to those practiced by commercial and enterprise organizations with respect to their customers' data. Organizations, for example, have adopted data log anonymization techniques, starting from the premise that the natural first solution to publish critical, privacy-preserving data is de-identification. In this technique, for the critical data set, removal of key identities from the records is performed. In combination with an external database, there are some other attributes to be used to identify the personal details, called Quasi-Identifiers.

Innovative Artifacts from a Technological and Commercial Perspective

Danneels (2002) defined a typology for product creation as of the combination of the status of two dimensions: (a) organizational mastery of the technologies involved, and (b) organizational mastery of customer demands (we will refer to this point in the text as commercial demand). Combining the alternatives "high domain" and "low domain" for these two dimensions, Danneels (2002) defined four types of actions for the generation of new products: pure exploitation, pure exploration, leveraging customer competence, and leveraging technological competence. Extrapolating this model to the context of scientific artifact creation, Gregor and Hevner (2013, 2014) defined four types of artifact creation: exploitation (or routine design), exploration, improvement, and exaptation.

According to Danneels (2002), when the organization widely dominates the technological and commercial dimensions of the new product, there is an action of the "pure exploitation" type. Exploitation type creations present a greater chance of success due to the organization's greater domination in relation to both dimensions, which reflect in projects with a well-defined method and objective (Turner & Cochrane, 1993). When bringing Danneels' model to the context of the creation of scientific artifacts within the DSR approach, Gregor and Hevner (2013, 2014) defined this type of creation initially as routine design and later as exploitation. Although for the business context the action of creating products from the action of exploitation is important, because it presents less risk and good capacity for revenue generation, for the context of scientific innovation its value is derisory. The science, represented by editors and reviewers of journals crave the new, artifacts that innovate in at least one of the dimensions, technological or commercial, which does not occur in artifacts generated from the action of exploitation (Goes, 2014).

Danneels (2002) defined the exploration action as the opposite to the exploitation action. Exploration actions present innovations to the company both in terms of technology and customer service. Turner and Cochrane (1993) emphasize that exploration-type projects present greater chances of failure due to greater difficulties in defining the method to be employed, as well as the objective. Because of the two innovation fronts and the high risk, "the returns from exploration are unclear, uncertain, and remote in time" (Danneels, 2002, p. 1106). Gregor and Hevner (2013) indicate significant scientific returns, but these situations are rarer to occur due to the doubly innovative demand, both from the technological and commercial dimensions.

A third product creation action addressed by Danneels (2002) is the leveraging technological competence. In this type of creation, the product or service continues to perform the same function but more efficient and effective (Tidd et al., 2005). According to Gregor and Hevner (2013) the favorable context for improvement action is when the company presents a product or service in a context in which the company has broad domain of the commercial dimension ("application domain maturity") and low domain of the technological dimension ("solution maturity"). The method of creation of improvement type implies in incurring technological risks, since it is a new technological proposition for a commercial problem already in the company's broad domain.

A fourth product creation action, opposite to the leveraging technological competence action, is the leveraging customer competence action (Danneels, 2002). Gregor and Hevner (2013) named this action as exaptation, characterized by high domain of the technological dimension ("solution maturity") and low domain of the commercial dimension ("application domain maturity"). In exaptation there occurs the opposite situation to the one observed in the method of creation by adaptation, in which it is observed new forms for old functions; in exaptation it is observed new functions for old forms (Dew & Sarasvathy, 2016). The most widespread example is the drug Viagra. Initially marketed for the treatment of pulmonary arterial hypertension (under the name Revatio) and later identified as suitable for the treatment of erectile dysfunction (Viagra). The same active ingredient was originally applied to cardiological treatment and later to the treatment of urological issues

(Dew & Sarasvathy, 2016). Thus, in the innovation of the exaptation type it is observed technological continuity and functional discontinuity (Andriani & Carignani, 2014). Thus, the creation of the exaptation type implies in incurring commercial risks, because it is about new commercial applications for technologies of broad domain of the company.

The analysis of the artifacts within the perspective of technological and commercial dimensions (Danneels, 2002; Gregor & Hevner, 2013, 2014) characterized there are three options of valid creative logics for a scientific artifact: invention, exaptation and improvement. The innovative logics most commonly found in DSR articles are those arising from creative actions of the improvement and exaptation type (Gregor & Hevner, 2014). This is due to the restrictions of the other two logics. The routine design or exploitation logic does not bring novelties, it is configured as more of the same and, therefore, without value within the scientific perspective. The invention logic is the most desired, but requires a larger set of innovative perceptions, of creative capacity, resulting in greater complexity, greater risk and intellectual challenge. In addition to being rarer due to the higher intellectual demand, there are also limitations or the cognitive biases of those who evaluate scientific publications. One of the biases is the greater receptivity and understanding from the anchoring in what is already known and domain of the reviewers; when much is innovative, as occurs in invention, the perception of quality may be harmed. From the editors' perspective, there is the distrust in relation to authors of scientific texts from regions or institutions without tradition in the scientific scenario (Lee et al., 2013). In summary, publishing articles with inventive artifacts implies overcoming many difficulties.

Obviously there are other terms or even variations of creative logics besides invention, exaptation, and improvement, which are employed for the development of scientific artifacts within the DSR approach. In the next subsection, we will explore creative logics from the perspective of researchers in the fields of innovation management and entrepreneurship. There is a broad set of terms used by researchers from these fields that are also employed for the characterization of artifacts developed with the DSR approach. Ontologically it is an important activity, considering that many of these terms, although distinct, have not only differences, but also similar characteristics.

CREATIVE TACTICS USED IN THE CREATION OF ARTIFACTS

A DSR artifact must have functions that solve a problem of practitioners or organizations that must be well delimited and characterized in the form of a "problem space." Depending on the problem, the solution may require the use of several basic heuristics. Savransky (2000, p. 13) identified 10 basic heuristics used for the problem-solving: neology, adaptation, multiplication, differentiation, integration, inversion, pulsation, dynamization, analogy, and idealization (a brief description of each heuristic is presented in Annex A). Recently, we analyzed the set of basic heuristics employed in the creation of 180 products and services, conceived and marketed by 51 entrepreneurs, small business owners who have been operating their businesses for more than 42 months (De Sordi et al., 2020). The various creation tactics or mechanisms practiced by these entrepreneurs in developing their products and services are described in Fig. 2.2. In the literature on innovation there is a set of terms linked to the practices of creation of products and services, very similar to the heuristic terms. This is entirely coherent, considering that this is a search for a solution to a very important problem for every entrepreneur: the creation and continuous renewal of the company's product and service portfolio.

It is observed that the four actions or logics described in the typology of Gregor and Hevner (2013, 2014) for the development of DSR artifacts are equally present in Fig. 2.2. The actions exaptation and improvement appear with the same denomination, the routine design action is analogous to the copy tactic, and the invention action is analogous to the new product development tactic. In addition to these four tactics directly associated with Gregor and Hevner's (2013, 2014) four types of descriptive logics, Fig. 2.2 presents seven other creative tactics: adaptation, custom-made, degradation, frugal innovation, new combination, nonaptation junk, and nonaptation spandrels. Although not mentioned in Gregor and Hevner's (2013, 2014) typology, these creative tactics are also useful to the DSR artifact creation process.

It is important to remember that complex logical activities usually involve more than one type of logical reasoning: abduction, inference, and deduction. As an example, DSR initially employs abductive reasoning for the idealization phase of possible solutions for the problem, subsequently, it employs deductive reasoning to evaluate the usefulness of the artifact and, finally, inductive reasoning to generalize the best solution

Adaptation. Unlike exaptation [see below], where function follows form, in adaptation a form follows function. In other words, the form of the entity is altered to deliver the same function to customers with the same demand (Andriani and Cattani, 2106). In adaptation, the focus is on the pre-existing fitness function (Andriani, Ali, and Mastrogiorgio, 2015). According to Dew, Sarasvathy, and Venkataraman (2004, p. 72) adaptation "is actually better described by the term "aptation" rather than adaptation, since the etymology of "aptus" is "fit", whereas "adaptus" refers to the process of increasing fitness by designing for a particular function".

Copy. Entrepreneurial copycats are those that begin a business with routines, competences, products and services that vary minimally in a particular market (Koellinger, 2008). Copying can occur in different ways, for example, by applying reverse engineering.

Custom-made. The custom-made mechanism involves a product or service that is developed specifically for a particular person or organization. The advent of the Internet of Things (IoT) permits the customization of many products or services that heretofore were mass produced and standardized (Desouza et al., 2007). Predictive technologies also facilitate this mechanism by using client data. A predictive technology is a body of tools capable of discovering and analyzing patterns in data so that past behavior can be used to forecast likely future behavior.

Degradation. The degradation mechanism involves the limitation of important features or the inclusion of unfavorable features. From the perspective of the product development function it is a step backwards in terms of the quality of the product or service. (Gershoff et al., 2012)

Exaptation. The Latin prefix "ex", meaning moving away from, exiting, extracting, which before the word aptation comes to indicate that the entity is shifting from its original purpose to perform another role in a new context. Therefore, in an exaptation kind of innovation technological continuity and functional discontinuity are observed (Andriani and Carignani, 2014). This is the opposite of what is observed in adaptation type methods, in which new forms perform old functions. In exaptation, new functions for old forms are observed (Dew and Sarasvathy, 2016).

Frugal innovation. Act of excluding non-essential functions that existed in the original product or service and/or replacing parts with cheaper analogous parts (Sarkar, 2011) to meet the needs of consumers with low buying power in a previously discriminated and neglected market (Zeschky, Widenmayer and Gassmann, 2011).

Improvement. Actions of improvement are intended to make something that already exists better and basically continue doing what it does, only more efficiently, more quickly, at a lower cost, better functions and more accurately, ... (Tidd, Bessant and Pavitt, 2005).

New Combination. A subclass of NPD is new combination, as proposed by Villani, Bonacini, Ferrari, Serra, and Lane (2007). In this way of framing NPD, the new product or service results from the integration of technological elements or processes that already existed previously. However, they were originally conceived and developed by different entrepreneurs. In this creation method, the entrepreneur operates more as an integrator of already existing entities.

New Product Development (NPD). This method involves activities distributed over three stages: predevelopment stage, development stage, and commercialization stage (Langerak, Hultink, and Robben, 2007). The resulting product or service "can be new to the business, new to the market, or new to the world". In other words, there is no need to be something new to the market and the world (Najafi-Tavani, Sharifi, Soleimanof, and Najmi, 2013, p. 3397). The aspect to highlight is that NPD results in something new at least to the one who develops it, in the context of this study, the entrepreneur.

Nonaptation Junk. Also known as technological nonaption, it is characterized as unexpected results obtained through technological research, often associated with new product development (NPD). Innovations based on nonaptation junk make use of inputs described as "materials and knowledge that just lies around" (Garud, Gehman, and Giuliani, 2016, p. 152). The best-known case is the experiment conducted at 3M that resulted in a "glue that did not glue", which was later recovered and constituted the Post-it note.

Nonaptation Spandrels. Creation using a resource hitherto without use, in other words not (from the Latin *non*) or without aptation. Gould (1997) adopted the term spandrel from architecture to illustrate how an empty space, initially without use in buildings, was later filled with mosaics and paintings and came to have an important aesthetic function in buildings (Villani et al., 2007). The prototypical example presented by Gould (1997) is Saint Mark's Cathedral in Venice, whose domes were built three centuries before the mosaics that now adorn them (Garud et al., 2016).

Fig. 2.2 Creative tactics used for the generation of new products and services (*Source* Adapted from De Sordi et al. [2019])

design for that class of problem (Kuechler & Vaishnavi, 2008). Similarly, it is observed in the innovation literature an interdependence between the various tactics or mechanisms of creation. For example, it is known that the exaptation mechanism is never an isolated creation mechanism. From the perception of functional alteration of a given entity, there is a need for actions to adapt it so that it can exercise a new function, in other words, exaptation being succeeded by an adaptation mechanism (Andriani & Carignani, 2014; Shumacher, 2012). We have then that the artifact conception in the DSR approach may also involve one or more creative tactics. Thus, it is prudent the conceptual domain of these various creative logics so that the researcher can better describe the process of creating his artifact.

In addition to terms from the innovation management area, some terms from the entrepreneurship area have been frequently employed in scientific texts that address artifacts developed with the DSR approach. They have been employed in the description of the artifact development process. Among these terms, three stand out: do it yourself (DIY), bricolage and effectuation. The most comprehensive term of the three is DIY. It is employed in function of the bipartite sense of DSR: pragmatic, while delivering an artifact for use; scientific, while delivering an artifact unique for being innovative in some sense, either technological and/or commercial. For being unique, justifying the adjective innovative attached to scientific knowledge, and for being something operational, characterizing its essential pragmatic, we have as the essence of the DSR approach the artifact developed by the researcher. Thus, all DSR has the delivery of something done by the researcher, i.e., DIY is in the essence of the action of the researcher who practices the DSR approach. Because it is a very broad term, we better work with the other terms already addressed, such as the types proposed by Gregor and Hevner (2013, 2014). Not only does it indicate that something has been done, but it already discriminates the type of innovation that occurred in relation to the technological and commercial dimensions.

The terms bricolage and effectuation are associated with the construction of something using whatever is at hand and available at the time of need. The central idea is the reduction of costs and risks from the resources already available. The pertinence with some of the creation mechanisms in Fig. 2.2 is in the use of resources already available to the developer. This occurs most explicitly with the exaptation, nonaptation junk, and nonaptation spandrel mechanisms. The incongruity of using the terms bricolage and effectuation for artifacts developed with the DSR approach is in the expectation of the efficiency of the artifact. While in DSR one expects superior performance of the new artifact over all others available, in bricolage and effectuation one only expects to meet the demand of the moment. This demand is usually associated with lack of time and/or resources of the one who needs the artifact for a certain function. Garud and his colleagues described bricolage as "an ad hoc quick fix that is ephemeral" that "it was not designed" and that "never became the basis for new commercial products" (Garud et al., 2016, p. 160). In this sense the bricolage and the effectuation are the extreme opposite of the creative logics employed in the DSR, which as indicated by the first letter of the acronym, value the Design of the new artifact.

References

Ahuja, G., & Lampert, C. M. (2001). Entrepreneurship in the large corporation: A longitudinal study of how established firms create breakthrough inventions. *Strategic Management Journal, 22*, 521–543.

Andriani, P., & Carignani, G. (2014). Modular exaptation: A missing link in the synthesis of artificial form. *Research Policy, 43*(9), 1608–1620.

Collins. (2013). *English dictionary.* http://www.collinsdictionary.com/dictio nary/english/cross-reference.

Corbin, J., & Strauss, A. (1990). Grounded theory research: Procedures, canons, and evaluative criteria. *Qualitative Sociology, 13*, 3–21.

Creswell, J. W. (2014). *Research design: Qualitative, quantitative, and mixed methods approaches* (4th ed.). Sage.

Danneels, E. (2002). The dynamics of product innovation and firm competencies. *Strategic Management Journal, 23*(12), 1095–1121.

De Sordi, J. O., Meireles, M., & de Oliveira, O. L. (2016). The Text Matrix as a tool to increase the cohesion of extensive texts. *Journal of the Association for Information Science and Technology, 67*(4), 900–914.

De Sordi, J. O., Nelson, R. E., Meireles, M., Hashimoto, M., & Junior, M. D. F. C. (2020). A longitudinal study of the creation methods used by entrepreneurs to develop new products and services. *International Journal of Entrepreneurship and Innovation Management, 24*(6), 482–502.

De Sordi, J. O., Nelson, R. E., Meireles, M., Hashimoto, M., & Rigato, C. (2019). Exaptation in management: Beyond technological innovations. *European Business Review, 31*, 64–91.

Dew, N., & Sarasvathy, S. D. (2016). Exaptation and niche construction: Behavioral insights for an evolutionary theory. *Industrial and Corporate Change, 25*(1), 167–179.

Eco, U. (1989). *Come si Fa una Tesi di Laurea*. Milano, Italy: Bompiani.

Garud, R., Gehman, J., & Giuliani, A. P. (2016). Technological exaptation: A narrative approach. *Industrial and Corprate Change, 25*, 149–166.

Goes, P. B. (2014). Design science research in top information systems journals. *MIS Quarterly, 38*, iii–viii.

Gregor, S., & Hevner, A. (2013). Positioning and presenting design science research for maximum impact. *MIS Quarterly, 37*(2), 337–355.

Gregor, S., & Hevner, A. (2014). The Knowledge Innovation Matrix (KIM): A clarifying lens for innovation. *Informing Science: THe International Journal of an Emerging Transdiscipline, 17*, 217–239.

Hatchuel, A., & Weil, B. (2009). C-K design theory: Na advanced formulation. *Research in Engineering Design, 19*, 181–192.

Hevner, A. R., March, S. T., Park, J., & Ram, S. (2004). Design science in information systems research. *MIS Quarterly, 28*, 75–105.

Jurczak, J. (2008). Intellectual capital measurement methods. *Economics and Organization of Enterprise, 1*(1), 37–45.

Kuechler, B., & Vaishnavi, V. (2008). On theory development in design science research: Anatomy of a research project. *European Journal of Information Systems, 17*(5), 489–504.

Lee, J. S., Pries-Heje, J., & Baskerville, R. (2011). *Theorizing in design science research*. Paper presented at the Proceedings of the 6th international conference on Service-oriented perspectives in design science research, Milwaukee, WI, USA.

Lee, C. J., Sugimoto, C. R., Zhang, G., & Cronin, B. (2013). Bias in peer review. *Journal of the American Society for Information Science & Technology, 64*(1), 2–17.

Newell, A., & Simon, H. (1972). *Human problem solving*. Prentice Hall.

Savransky, S. (2000). *Engineering of creativity: Introduction to TRIZ methodology of inventive problem solving*. Boca Raton, FL: CRC Press.

Shumacher, P. (2012). *The autopoiesis of architecture: A new agenda for architecture* (Vol. 2). Wiley.

Simon, H. A. (1976). The business school: A problem in organizational design. In H. A. Simon (Ed.), *Administrative behavior: A study of decision-making processes in administrative organization* (pp. 335–356). New York: Free Press.

Simon, H. A. (1996). *The sciences of the artificial* (3rd ed.). MIT Press.

Taura, T., & Nagai, Y. (2012). *Concept generation for design creativity: A systematized theory and methodology*. London: Springer.

Tidd, J., Bessant, J., & Pavitt, K. (2005). *Managing innovation: Integrating technological, market and organizational change* (3rd ed.). Chichester, UK: John Wiley & Sons.

Turner, J. R., & Cochrane, R. A. (1993). Goals-and-methods matrix: Coping with projects with ill defined goals and/or methods of achieving them. *International Journal of Project Management, 11*(2), 93–102.

van Aken, J. E., & Romme, G. (2009). Reinventing the future: Adding design science to the repertoire of organization and management studies. *Organization Management Journal, 6*(1), 2–12.

van De Ven, A. H. (2007). *Engaged scholarship: A guide for organizational and social research.* Oxford University Press.

Wasserman, S., & Faust, K. (1994). *Social network analysis: Methods and applications.* Cambridge University Press.

Yin, R. K. (2014). *Case study research: Design and methods* (5th ed.). Sage.

CHAPTER 3

Types of Artifacts or Knowledge Generated by DSR

According to Hevner et al. (2004, p. 77) the scientific artifact developed from the DSR method may have one or more of the following characteristics: "constructs (vocabulary and symbols), models (abstractions and representations), methods (algorithms and practices), and instantiations (implemented and prototype systems)." It is important to note that this description was developed for the context of the information systems area. The literature on the types of artifacts is not conclusive about a taxonomy for artifacts, considering the many peculiarities of the various areas of science. In this chapter we will explain the four most widely spread types of artifacts, which are the four described by Hevner et al. (2004). We will also explore the dynamics and relationships between these artifacts from published DSR.

CONSTRUCT

According to the Merrian-Webster dictionary the construct is "something constructed by the mind: such as a theoretical entity or a working hypothesis or concept." For the philosophy of science the construct is an ideal object whose existence depends on the human mind. Intrinsically, the construct must contribute to the community of practitioners by attributing a semantic value, being this characteristic that gives it meaning and value. Bunge (1974, p. 1) highlights that "interpretation of

J. O. De Sordi, *Design Science Research Methodology*,
https://doi.org/10.1007/978-3-030-82156-2_3

constructs is performed by semantic assumptions." Bringing the understanding of the physical or abstract entity to practitioners, beneficiaries and other actors involved with the artifact is relevant in order to enable the dialogue and the understanding of the entity that is a constituent part of the artifact and, consequently, of the solution.

Scientific artifacts developed according to the proposal of the DSR approach encompass scientific knowledge from different areas, specializations and sub-specializations. Thus the intertextual composition can bring together terms from distant areas, from different contexts, as well as constitute new entities to be conceptualized. As a recent example, already discussed in this text, of the incorporation of terms from other areas in an innovative way in another area, we have the use of the term exaptation. Initially developed by paleontologists to explain the evolution of species, it was later applied to discuss the process of innovation in products and services. The first authors, when incorporating this concept to the theories of innovation management, had to work and develop the construct expatation for this specific context. Another example of an innovative and widespread construct, from the beginning of the twentieth century, is the Gantt Chart. By means of graphical symbols, overlapping horizontal bars, this artifact allowed to graphically demonstrate the dependency between the activities of a project as well as the time duration of each one of them and of the project as a whole. In this case we have an artifact of construct type constituted not by vocabularies, but by symbols.

The constructs give meaning to important concepts so that it can occur actions of use of the artifact by practitioners, either through the semantic value of their vocabularies or their symbols. In promoting the work developed by practitioners, artifacts of construct type demonstrate their usefulness. Here a series of analogies can be made as artifacts of our everyday life, of collective use and understanding. Among the artifacts of the construct type based on symbols, we can think of the functionality and usefulness of graphic signs (signs) instituted to guide vehicular and pedestrian traffic. Another example is the creation of Arabic numerals for mathematical operations. Among the artifacts of the construct type based on vocabulary, we can think of the elements necessary for understanding the formula of speed. The actors involved in the conception of the artifact to calculate the velocity had to define and have domain of words such as displacement in space and time interval.

There are examples of constructs used even in the development of the DSR approach itself. To become an approach as an artifact, in the form of scientific method as we know it today, there was a series of necessary developments for the DSR, including its constructs. A good example is the very definition of the term Design. In the beginning of the last decade, McKay et al. (2012) evidenced that the concept of design for professionals in the area of information systems was much more limited conceptually than the understanding of practitioners in other areas. The perception of design restricted to three concepts (problem solving, artifact production, and process) was expanded to ten concepts. The concepts of: intention; planning (including modeling and representation); communication; user experience; value; professional practice; and service were included. These new concepts incorporated a more humanistic perspective to the design from the perspective of information systems professionals who began to value and pay more attention to the demands of the various actors involved with the artifact.

Our research in repositories of scientific articles showed that the type of artifact less published is the construct type, we will present more details of this research in the last section of this chapter. An even more difficult aspect is to find a DSR whose result, the artifact, is exclusively of construct type. Table 3.1 presents, as an example, some recent DSR articles that published exclusively construct artifacts. It is noteworthy that the artifacts are proposed to practitioners from three different areas.

MODEL

"A model is a set of propositions or statements expressing relationships among constructs. In design activities, models represent situations as problem and solution statements" (March & Smith, 1995, p. 256). The interpretation of this statement is important to value the constructs. The construct can be the novel aspect itself, as discussed in the previous subsection, or they can be the enabler or means of transmitting the novel aspect in the form of a model. As an enabler of the new artifact, the construct is useful in making the new model explicit. To illustrate this situation imagine that we are in the mid-1990s and we are making explicit the new model for organizational purchasing in the form of a reverse auction. For this, we represent and describe the various actors, information flows, and interactions between them, rules, activities, products, and data from other entities necessary to operationalize the reverse auction environment. For this, we use words such as process, state, actor, data flow, interaction, decision, internal and external products, among other terms. These

Table 3.1   Examples of construct artifacts

Application area:	Engineering (project development)
Artifact name:	Study of the challenges of collaborative design
Utility:	"We lay a ground for design and development of better support for collaborative design"
Source:	Piirainen, K. A., Kolfschoten, G. L., & Lukosch, S. (2012). The joint struggle of complex engineering: A study of the challenges of collaborative design. *International Journal of Information Technology & Decision Making, 11*(6), 1087–1125
Application area:	Corporate governance
Artifact name:	IR-3D conceptual framework
Utility:	"The taxonomy structure (IR-3D) is targeted at entities interested in designing an XBRL taxonomy for integrated reporting (IR). This is a call for academics and practitioners to explore the potential of technology to improve corporate disclosure and open up new projections for resurging themes on intellectual capital (IC) reporting with prospects for IC "fourth-stage" research focused on IC disclosure."
Source:	La Torre, M., Valentinetti, D., Dumay, J., & Rea, M. A. (2018). Improving corporate disclosure through XBRL: An evidence-based taxonomy structure for integrated reporting. *Journal of Intellectual Capital, 19*(2), 338–366
Application area:	Information Systems (software specification)
Artifact name:	The Graphical Syntax for Actor–Network Theory
Utility:	"The utility of actor-network theory as a conceptual tool for information systems research can be increased by expressing actor-network theory in a graphical format. To this end, a graphical syntax was designed based on a comprehensive conceptualisation of actor-network theory."
Source:	Alexander, P. M. & Silvis, E. (2014). Towards extending actor-network theory with a graphical syntax for information systems research. *Information Research, 19*(2), paper 617

concepts are made explicit through symbols when we develop several diagrams, such as state transition, process decomposition, actor interaction, all of them well known and widespread since the 1980s. So, the new fact is not the constructs, but the content that they are transmitting, that is, the new innovative model of the reverse auction environment. In this new environment, who opens the purchase contest is not the selling agent as in the conventional auction, but the buyer; the bids do not grow, they decrease; among many other differences. Although there are similarities with the conventional auction procurement model, there are also many distinct aspects between the two models.

It makes perfect sense the first authors who developed the seminal texts of the DSR approach in presenting first the artifact of the construct type to then define the artifact of the model type, considering this logical sequence. Models are useful both to bring explicitness and insight to as yet unknown or poorly formulated problems and to present solutions to such problems (March & Smith, 1995). Mapping and clearly describing a problem is usually a great challenge or a great problem. In this category fall all new artifacts for diagnosis of human health parameters. For example, the artifact that generates the graphical representation of a person's brain activity is not the solution to the disease, but it is part of the solution to the initial problem, that of helping to diagnose the causes and identify the disease to be treated. Thus, both descriptive problem models and solution presentation models result in useful artifacts to practitioners, whose dynamic between them is well characterized by Hevner et al. (2004, p. 78): "Models aid problem and solution understanding and frequently represent the connection between problem and solution components enabling exploration of the effects of design decisions and changes in the real world."

In a broad sense, the word model is a description, an analogy, an example for imitation, that is, it is a metaphor of how something is or how it should be. Therefore it is not something broad, complete, and totally correct, it is characterized as a reduction of the entity it represents. Despite these limitations it is something useful to the purposes for which it is intended. Virtually all the development of new complex artifacts are previously designed through models, whether they are vehicles, buildings, software, and other complex human developments. Table 3.2 presents, as an example, three DSR artifacts of the model type that have been recently published and that are directed to practitioners in three different areas.

METHOD

The most commonly cited method artifacts are the algorithms and practices (Hevner et al., 2004; March & Smith, 1995). As examples of algorithms the simplest and most direct example are the instructions for computer languages that result in software or sets of integrated software in the form of information systems. The practices are similar in the sense that they characterize the way a certain activity should be performed. In the organizational environment, many of these most coveted working methods, in terms of being copied or integrated into organizations, are

Table 3.2 Examples of model artifacts

Application area:	Management (value creation)
Artifact name:	Value Analysis Model (VAM)
Utility:	"VAM is proposed to measure the value creation expected by customers and to identify value losses through indexes"
Source:	Giménez, Z., Mourgues, C., Alarcón, L. F., Mesa, H., & Pellicer, E. (2020). Value analysis model to support the building design process. *Sustainability, 12*(10), 4224
Application area:	Engineering (construction project)
Artifact name:	Model to design modular buildings
Utility:	"The model provides a hands-on tool for practitioners to design modular buildings."
Source:	Gravina da Rocha, C., El Ghoz, H. B. C., & Jr Guadanhim, S. (2020). A model for implementing product modularity in buildings design. *Engineering Construction & Architectural Management, 27*(3), 680–699
Application area:	Healthcare predictive analytics
Artifact name:	Bayesian Multitask Learning (BMTL)
Utility:	"BMTL allows healthcare providers to achieve multifaceted risk profiling and model an arbitrary number of events simultaneously."
Source:	Lin, Y.-K., Chen, H., Brown, R. A., Li, S.-H., & Yang, H.-J. (2017). Healthcare predictive analytics for risk profiling in chronic care: A Bayesian multitask learning approach. *MIS Quarterly, 41*(2), 473-A3

called best practices. Although not an adequately scientifically defined term, the term best practices serves as a facilitator for communication between researchers and practioners during the development of DSR projects (Burgoyne & James, 2006, p. 307):

> [...] the researchers were troubled by the notion of 'best practice', which was difficult to define. With the brief agreed, it seemed more 'obvious' to the practitioners what constituted 'best practice' than to the researchers. The working group explicitly stated their belief that 'an organisation that has been in the top league for some time must be doing something right and is worth studying'.

Although the method-type artifact does not need to be linked to a model-type artifact, March and Smith (1995, p. 257) indicated quite clearly the link between these two types of artifacts: "Methods can be tied to particular models in that the steps take parts of the model as input. Further, methods are often used to translate from one model or representation to another in the course of solving a problem." As already observed in the transition of presentation between the artifacts of construct type to the model type, the main authors of DSR also present the artifacts of method type succeeding the artifact of model type. Although the most common sequencing of creative activities for the development of artifacts is from model type to method type, the inverse path in terms of support for the creation of new artifacts can also occur: "the desire to use a certain type of method can influence the constructs and models developed for a task" (March & Smith, 1995, p. 257).

While the artifact type model is somewhat broad and open as to its conception, scope and way of reading, the artifacts of the method type are more specific and restricted. Let's take as an example the reading of artifacts of these two types. A model artifact can be understood by reading any of its components, regardless of the order. When reading a model-type artifact that describes an innovative data modeling described in an Entity-relationship Model, we can read it from any of its different entities, that is, from different perspectives. When studying the model for organizational purchasing in the form of reverse auction we can start the study and understanding by the actors involved and the interaction between them, or by the main processes, or by the main data entities involved, i.e., it is a broad conceptual model, which does not imply, for example, in a specific reading order. The same does not occur with the method-type artifact, which is much more punctual and its reading order should be respected, and should be the vertical or sequential reading.

Methods define processes and considering that the organizations in which we work are large collections of processes (Dijkman et al., 2012), we have that the artifact type method presents a context of application and understanding by practitioners much greater than those associated with the artifacts of type construct and model. As we will observe in the last subsection of this chapter, the artifacts of the method type are found in the scientific literature developed with the DSR approach in a quantity higher than the other two types of artifacts. Table 3.3 presents, as an example, three DSR artifacts of the method type that were recently published and that are directed to practitioners from three different areas.

Instantiation

The instantiation type artifacts are defined as: physical implementations (Pries-Heje & Baskerville, 2008); implemented and prototype systems, a type of system solution (Hevner et al., 2004); and "the realization of an

Table 3.3   Examples of method artifacts

Application area:	Information systems (database forensic)
Artifact name:	Common Database Forensic Investigation Processes (CDBFIP)
Utility:	"The artifact uses database content and metadata to reveal malicious activities on database systems in an Internet of Things, proposing a common investigation process for database forensics"
Source:	Al-Dhaqm, A., Razak, S., Othman, S. H., Choo, K.-K. R., Glisson, W. B., Ali, A., & Abrar, M. (2017). CDBFIP: Common database forensic investigation processes for internet of things. *IEEE Access*, 5, 24401–24416
Application area:	Supplier Management (logistics risk)
Artifact name:	Supply Chain Sustainability Risks (SCSR)
Utility:	"Procedural model that facilitates an identification of SCSR 'hotspots'"
Source:	Busse, C., Schleper, M. C., Weilenmann, J., & Wagner, S. M. (2017). Extending the supply chain visibility boundary. *International Journal of Physical Distribution & Logistics Management*, 47(1), 18–40
Application area:	Environment (performance indicators)
Artifact name:	Carbon Tracker
Utility:	"Artifact provides logistics companies with enriched primary greenhouse gas emissions data in an automated and timely fashion and with analysis features that support organizational sensemaking with regard to eco-sustainability"
Source:	Hilpert, H., Kranz, J., & Schumann, M. (2013). Leveraging Green IS in logistics. *Business & Information Systems Engineering*, 5(5), 315–325

artifact in its environment" (March & Smith, 1995, p. 258). The central idea to be highlighted is the demonstration in fact of the usefulness, of the effectiveness of the proposed artifact, whether it is a construct, a model or a method. Here it does not matter if it is a well evolved version of the artifact or if it is an initial version, the important thing is that this version of the artifact is operational and that the practitioner can use it. The core value to be highlighted in instantiation is the practical analysis of the viability and usefulness of the artifact for solving the problem space it proposes to address. With the instantiation of the artifact, the DSR approach allows verifying if the result of using the artifact can be considered an effective solution for that type of problem.

The definition of March and Smith (1995) is interesting because it explains that the execution and evaluation of the effectiveness of the artifact should occur in the natural environment of use of the artifact, considering the impact on practitioners and beneficiaries of the action of the artifact. The great benefit of the scientific artifact developed with the DSR approach materializes, is proven by its instantiation. It is by the instantiation of the artifact that occurs the identification of subtle deficiencies, difficult to be identified in the initial versions of the artifact design without the occurrence of its effective use. The identified deficiencies are corrected, a new version of the artifact is generated and operationalized by the practitioners (field tests). Thus the process restarts with the identification of new demands, new adjustments of the artifact, new versions, new operationalizations, closing a virtuous cycle of learning the best design by the continuous practice of use. Instead of the term *learning by doing, the* correct term here is *learning by designing* (Rohde et al., 2017).

Models that describe the DSR approach in detail, in terms of its phases and activities, such as the one by Peffers et al. (2007), clearly differentiate the artifact demonstration phase from the artifact evaluation phase. This distinction highlights that a complete proposition, following the scientific rigors of the DSR approach, should go beyond a demonstration. There are few researches or projects with development of DSR artifacts that manage to propose a new artifact and make its evaluation in only one article. Most complete DSR projects, ranging from the initial project to its effective evaluation, are portrayed in two or more articles. Here is a reflection: should these articles addressing the first phases of DSR artifacts, which did not go through all phases of the DSR method (not presenting, for example, the evaluation phase) indicate the DSR approach as the driver of their research actions? I think so, as long as there are texts that point out these deficiencies, in the form of limitations of the research and proposals for continuing the research. The model of Peffers et al. (2007) itself presents entry points that allow the resumption of a DSR at various times, four of them before the evaluation phase.

The fragmentation of the DSR in several articles occurs in practice, considering that most researches with artifacts published with the designation of DSR approach did not conduct and/or do not present data from the evaluation phase. This is perceived even in DSR articles published in high impact journals. We should understand this within a normality, considering the historical context of evolution of scientific knowledge itself in an integrated, continuous and collaborative way. This situation should be understood as a research opportunity to be worked on by other researchers. As there is a natural repulsion of editors for publishing unsuccessful results, it would be interesting that every journal that publishes

DSR artifacts without proper evaluation of the same that assumed the commitment to open space for complementary articles, including those that applying the methodological rigor did not find the utility initially claimed by the artifact. With this, it would increase the interest of other researchers in accessing or reproducing the artifact already published with the aim of conducting the evaluation of the same.

The main DSR canons, especially those linked to applied knowledge, such as those that indicate the reduction of time between invention and innovation, besides the delivery of useful knowledge for "ready consumption," only occur after the effective instantiation of the artifact. Thus, a successful natural evolution process for the other three DSR artifact types (construct, model and method) involves the generation of an instantiation type artifact. More pragmatically, we should have the instantiation artifact type as a necessary condition for a DSR project to be considered complete and successful. If the artifact in question will become a marketed product or service is another challenge, involving other demands that are not always in the purview of researchers. However, ensuring that the testing of the DSR artifact occurs in the field, being used by typical practitioners, in their natural working environment, and with inputs (data and/or raw material) from the real-life environment, configures itself as a legitimate demand for the researcher who practices the DSR approach.

Thus, a DSR that proposes an artifact only ends and can only be considered effectively concluded when at some point a team publishes a scientific text with its evaluation. When conducting a longitudinal research with the historical regression of DSR articles that effectively performed the evaluation phase, we noticed that there is a trajectory composed of several previous articles that substantiated the published artifact. Many of these predecessor articles only defined the proposal of the artifact (construct, model and/or method), others present and demonstrate, while others only demonstrate the artifact already presented previously. We want to highlight that there are many trails, possible paths to be taken by researches that develop artifacts with the DSR approach. Although the ideal would be a broad and full-bodied article, covering from the proposition of the artifact to its broad and effective evaluation, this is not always possible and generally this is not what occurs in publications of articles that make use of the DSR approach.

The information technology industry in conjunction with large consulting firms is prodigious in proposing business solutions as of the integration and combination of already existing technologies, that is,

Table 3.4 Examples of instantiation artifacts

Application area:	Pharmacy (patient experience)
Artifact name:	ePharmacare
Utility:	"Address the challenges of community pharmacists' integration with primary healthcare services."
Source:	Lapão, L. V., da Silva, M. M., & Gregório, J. (2017). Implementing an online pharmaceutical service using design science research. *BMC Medical Informatics and Decision Making, 17*(1), 1–14
Application area:	Civil Engineering (project consultation)
Artifact name:	Mobile BIM AR System (Building Information Modelling (BIM) and Augmented Reality (AR))
Utility:	"Improving the information retrieval process during construction"
Source:	Chu, M., Matthews, J., & Love, P. E. D. (2018). Integrating mobile building information modelling and augmented reality systems: An experimental study. *Automation in Construction, 85*, 305–316
Application area:	People Administration (agenda management)
Artifact name:	Online Calendar Services (OCS)
Utility:	"Distribute collaborative work that spans temporal, geographic, thematic and digitally inscribed boundaries thereby re-aligning organizational routines"
Source:	Akoumianakis, D., & Ktistakis, G. (2017). Digital calendars for flexible organizational routines. *Journal of Enterprise Information Management, 30*(3), 476–502

applying creative tactics of the new combination type (Villani et al., 2007). Many evaluations of these solutions proposed by the "market" are developed using the DSR approach, working notably the instantiation type artifact, considering that the solution is already configured and presented by the proponent organizations. Table 3.4 presents, as an example, three instantiation type artifacts that have been recently published and that are directed to practitioners from three different areas.

EVOLUTIONARY CYCLES OF DSR ACCORDING TO THE DIFFERENT TYPES OF ARTIFACTS

In 2016, while attending an international congress in the area of Administration, I attended a presentation that indicated that only 2% of all "recommendations" to managers published in the form of scientific article had been tested. The very authors who proposed the theory did not test it, those who should be the most interested in pursuing the research. From this information we decided to verify the deployment of the artifacts published with the DSR approach. Thus, in August 2018, I and some colleagues from the research team conducted a search in the Web of Science (WoS) repository with the purpose of identifying scientific articles associated with the DSR approach. For the search command, scanning the WoS database, we employed the following criteria: (a) for the "Topic" attribute we used two terms (("Design Science") and ("Artifact" or "Artefact")); and (b) for the "Document Type" attribute we used the term "Article." This search resulted in the identification of 156 articles that, after an initial analysis through skimming type reading (Duggan and Payne, 2009), resulted in 92 articles associated with applied research with the DSR approach. We discarded the theoretical articles, focused on the dissemination of the DSR approach, as well as those that presented variations of the approach from triangulation with the DSR approach.

By applying the content analysis technique to the corpus of texts of the 92 articles in our sample, we identified the name, function and type for each of the proposed artifacts. From this analysis, we identified a total of 136 artifacts developed and presented by these 92 articles. In classifying these artifacts by type, we obtained: 19 artifacts of the construct type, 41 of the model type, 41 of the method type, and 35 of the instantiation type. We observed the following frequencies of artifacts per article: 56 of the 92 articles in the sample (61%) developed and presented 1 artifact; 33 articles (36%) developed and presented 2 artifacts; 6 articles (7%) developed and presented 3 artifacts; and 1 article with no artifact. It is noteworthy, as expected, that the most frequent situation is the development and presentation of one artifact per article.

For the analysis of the trajectory of these 92 DSR we identified the articles published previously and subsequently and that were directly associated with the same logical construction, i.e., the development of the same technological artifact. For this, we adopted the following strategies:

a. analysis of the curricula vitae of the authors of each of the 92 articles, aiming to identify articles associated with the artifact in question;
b. analysis of the texts of the 92 articles aiming to identify previous constructions (predecessor articles) formally highlighted or that could be inferred from citations throughout the text;
c. search for the names and acronyms of the artifacts in repositories of scientific articles (WoS, Proquest, JStor, EBSCO), social networks

of researchers (ResearchGate) and the Internet in general (Google Scholar).

For the initially identified predecessor and successor articles the following procedures were carried out: (a) skimming type search to verify the relevance of the article with the artifact stated in the sample article; in case of relevance, (b) content analysis aiming to identify the type of association in terms of artifact sequencing. This process resulted in the identification of predecessor and successor articles associated with 57 of the 92 articles in the sample, i.e., 57 (62%) DSR applied researches with a history of two or more associated articles. As of the artifact types identified in the historical sequence of each of these 57 research projects, with more than one article and with at least one transition between artifacts, Table 3.5 was prepared to describe the transitions between artifact types for the 57 DSR projects associated with two or more publications in the form of articles.

The longitudinal analysis of the 57 DSR projects that were continued beyond one article allowed us to identify the types and frequencies of each of the possible development trajectories of research projects with a DSR approach. Nine opportunities for entry points were identified for continuing DSR:

   i. Enhance a construct artifact type, observed 1 occurrence (construct-construct);
   ii. Improve an artifact of type model, observed 8 occurrences (model-model);
   iii. Enhance an artifact of type method, observed 2 occurrences (method-method);

Table 3.5 Transitions between the types of artifacts observed in the articles associated with the 57 DSR projects analyzed

	Construct	*Model*	*Method*	*Instantiation*	*Outdegree*
Construct	1	8	13	10	32
Model	–	8	2	2	12
Method	–	–	2	2	4
Instantiation	–	–	–	–	0
Indegree	1	16	17	14	48

iv. From an artifact of type construct develop an artifact of type model, observed 8 occurrences (construct-model);

v. From an artifact of the construct type develop an artifact of the method type, observed 13 occurrences (construct-method);

vi. From an artifact of type model develop an artifact of type method, observed 2 occurrences (model-method);

vii. From an artifact of type construct develop an artifact of type instantiation, observed 10 occurrences (construct-instantiation);

viii. From an artifact of type model develop an artifact of type instantiation, observed 2 occurrences (model-instantiation);

ix. From an artifact of type method develop an artifact of type instantiation, observed 2 occurrences (method-instantiation).

The longitudinal perception of DSR, covering one or more artifacts described in one or more articles, brings some benefits to the scientific community. The first of these is to assist in the identification of research opportunities, having as starting point the last type of artifact published. The entry points presented by Peffers et al. (2007) are defined in the perspective of phases of the DSR approach, whereas the entry points presented in this subsection are in the perspective of types of artifacts of the DSR. Thus, the perspective of entry points presented by Peffers et al. (2007) in conjunction with the one presented in this subsection are added, giving a more comprehensive perspective to researchers who are interested in pursuing DSR. Awareness of the possible research directions from the types of DSR artifacts allows authors of research with the DSR approach to more easily identify and declare opportunities for further research. The article by Genemo et al. (2016, p. 11) is a good example of such a statement:

At this stage the development will concentrate on building constructs and model artefacts. Once the development, building and evaluation of the two artefacts are established, the research will progress to the next artefacts - methods and instantiations - development and building.

References

Bunge, M. A. (1974). *Treatise on basic philosophy volume 2: Semantics II- interpretation and truth.* Kluwer Academic Publishers.

Burgoyne, J., & James, K. T. (2006). Towards best or better practice in corporate leadership development: Operational issues in mode 2 and design science research. *British Journal of Management, 17*(4), 303–316.

Dijkman, R., Rosa, M. L., & Reijers, H. A. (2012). Managing large collections of business process models-current techniques and challenges. *Computers in Industry, 63*(2), 91–97.

Genemo, H., Miah, S. J., & McAndrew, A. (2016). A design science research methodology for developing a computer-aided assessment approach using method marking concept. *Education and Information Technologies, 21*(6), 1769–1784.

Hevner, A. R., et al. (2004). Design science in information systems research. *MIS Quarterly, 28,* 75–105.

March, S. T., & Smith, G. F. (1995). Design and natural science research on information technology. *Decision Support Systems, 15*(4), 251–266.

McKay, J., Marshall, P., & Hirschheim, R. (2012). The design construct in information systems design science. *Journal of Information Technology, 27*(2), 125–139.

Peffers, K., Tuunanen, T., Rothenberger, M. A., & Chatterjee, S. (2007). A design science research methodology for information systems research. *Journal of Management Information Systems, 24*(3), 45–77.

Pries-Heje, J., & Baskerville, R. (2008). The design theory Nexus. *MIS Quarterly, 32*(4), 731–755.

Rohde, M., Brödner, P., Stevens, G., Betz, M., & Wulf, V. (2017). Grounded Design—A praxeological IS research perspective. *Journal of Information Technology, 32*(2), 163–179.

CHAPTER 4

DSR from the Perspectives of Different Areas or Professional Schools

For further insight into the insertion of the DSR approach in research and in the different professional schools, we conducted a search in three repositories of scientific articles to identify the areas and journals that most publish and disseminate DSR approach. The search was conducted during the month of March 2021 at EBSCO, ProQuest, and PubMed repositories. The search algorithm was configured to present articles that presented the term "design science research" in the abstract. As a search criterion, it was also defined to present only scientific texts in the form of articles in English. To facilitate the manipulation and exposition of the data, we worked only with journals that published two or more articles on DSR. Table 4.1 presents the result of this search, indicating 340 articles, from 72 journals, associated with 8 distinct areas.

A quick analysis of the data in Table 4.1 indicates that the exponent area both in number of articles and in the diversity of journals is that of Information Systems. Next comes the area of Administration. We can compose several other categories for these articles and journals, considering the multifaceted nature of scientific artifacts, generally arising from the triangulation between different concepts and knowledge. For instance, we might have a category of journals focused on the theme decision-making, which would encompass 15 articles distributed in 4 journals: *Journal of Decision Systems*, *Decision Support Systems*, *Decision*

J. O. De Sordi, *Design Science Research Methodology*, https://doi.org/10.1007/978-3-030-82156-2_4

Table 4.1 Areas and journals that most publish articles with the DSR approach

Area	Journal name	Qty art	EBSCO	ProQuest	PubMed
Administration	Business Process Management Journal	9	X	X	
	International Journal of Accounting Information Systems	6	X		
	Journal of Decision Systems	6	X		
	Business Research	5	X	X	
	Decision Support Systems	5	X		
	Journal of Service Research	3	X		
	Transforming Government: People, Process and Policy	3		X	
	Benchmarking: An International Journal	2	X	X	
	Electronic Commerce Research & Applications	2	X		
	International Journal of Managing Projects in Business	2	X		
	International Journal of Physical Distribution & Logistics Management	2	X		
	Business Systems Research	2		X	
	Electronic Markets	2		X	
	Total Journals: 13 **Total Articles:**	**49**			
Construction	Engineering, Construction and Architectural Management	4	X	X	
	Construction Management & Economics	2	X		
	Lean Construction Journal	2	X		
	Total Journals: 3 **Total Articles:**	**8**			

(continued)

Table 4.1 (continued)

Area	Journal name	Qty art	EBSCO	ProQuest	PubMed
Education	Interactive Technology and Smart Education	3		X	
	Decision Sciences Journal of Innovative Education	2	X		
	Education and Information Technologies	2		X	
	International Journal of Education and Development Using Information and Communication Technology	2		X	
	Total Journals: 4 **Total Articles:**	9			
Engineering	International Journal of Production Research	7	X		
	Journal of Operations Management	3	X		
	Requirements Engineering	3		X	
	Brazilian Journal of Operations & Production Management	2	X		
	IEEE Transactions on Engineering Management	2	X		
	International Journal of Operations & Production Management	2	X	X	
	Central European Journal of Operations Research	2	X	X	
	South African Journal of Industrial Engineering	2		X	
	Total Journals: 8 **Total Articles:**	23			

(continued)

Table 4.1 (continued)

Area	Journal name	Qty art	EBSCO	ProQuest	PubMed
Environment	Sustainability	34	X	X	
	Journal of Cleaner Production	5	X		
	Built Environment Project and Asset Management	2		X	
	Total Journals: 3 **Total Articles:**	**41**			
Health	Studies in Health Technology and Informatics	7			X
	Journal of Medical Internet Research	5			X
	International Journal of Environmental Research and Public Health	3			X
	BMC Medical Informatics and Decision Making	2		X	X
	International Journal of Medical Informatics	2			X
	JMIR mHealth and Health	2			X
	Total Journals: 6 **Total Articles:**	**21**			
Information Management	Journal of Enterprise Information Management	7	X	X	
	Information & Management	6	X		
	Government Information Quarterly	3	X		
	International Journal of Information Management	3	X		
	Information Polity	2	X		

(continued)

Table 4.1 (continued)

Area	Journal name	Qty art	EBSCO	ProQuest	PubMed
	Electronic Journal of Knowledge Management	2		X	
	Information & Security	2		X	
	Information Management & Computer Security	2		X	
	Total Journals: 8 **Total Articles:**	27			
Information Systems	*European Journal of Information Systems*	21	X		
	Journal of the Association for Information Systems	15	X	X	
	Communications of the Association for Information Systems	15	X	X	
	Information Systems & e-Business Management	10	X		
	Information Systems Frontiers	10		X	
	Information Systems and eBusiness Management	9		X	
	Journal of Management Information Systems	8	X		
	MIS Quarterly	8	X		
	Business & Information Systems Engineering	8		X	
	Information Systems Journal	6	X		
	Information Technology for Development	5	X		
	Journal of Database Management	5	X		
	Australasian Journal of Information Systems	4	X		

(continued)

Table 4.1 (continued)

Area	Journal name	Qty art	EBSCO	ProQuest	PubMed
	AIS Transactions on Human–Computer Interactions	4		X	
	Information Technology & People	4		X	
	Computers & Security	3	X		
	Computers in Industry	3	X		
	Electronic Journal of Information Systems Evaluation	3	X		
	Information & Software Technology	3	X		
	Information Systems Research	3	X		
	Pacific Asia Journal of the Association for Information Systems	3		X	
	Future Generation Computer Systems	2	X		
	International Journal of Computer Integrated Manufacturing	2	X		
	Journal of Computer Information Systems	2	X		
	Systems Research & Behavioral Science	2	X		
	Empirical Software Engineering	2		X	
	Journal of Systems and Information Technology	2		X	
	Total Journals: 27 **Total Articles:**	162			

Sciences Journal of Innovative Education, and *Decision Sciences Journal of Innovative Education*.

Besides the area of Information Systems, whose professionals by natural vocation are directed to the development of technological artifacts, there are seven other areas, all recognized as professional schools. They clearly present groups of professionals with well-defined functions. The Administration area presents functional labels such as directors, managers, and analysts, all of whom focus on at least one of the company's several resources, such as marketing, strategy, production, finances, people, and quality, among others. The construction area has the architects and engineers responsible for designing and executing new constructions, as well as performing renovations of existing buildings. Also make extensive use of various artifacts in the exercise of their functions the professionals of the health area, such as doctors and nurses. From clinical analysis, decision-making regarding the most appropriate treatment, to the execution of procedures, there are many artifacts in use. In the Information Management area we have besides librarians and information scientists, information center analysts and other professionals who take care of the development and use of information collections in organizations. Professionals in the field of education also face several challenges that require specific artifacts. We highlight here only three very specific situations that demand very specific strategies and artifacts: the education of children with disabilities, the literacy of senior citizens, and the literacy of people in a language other than their mother tongue. There are also the growing challenges for professionals dealing with environmental issues, through to the widely publicized design and implementation challenges of engineering professionals. In short, we identified DSR articles associated with journals directed to areas whose professionals make intensive use of artifacts in the exercise of their functions.

LONGITUDINAL EVOLUTION OF THE ADOPTION OF THE DSR APPROACH BY RESEARCHERS

Table 4.2 presents the distribution by publication year of the DSR articles discussed in the previous section, all from the EBSCO, ProQuest and PubMed repositories. We can observe in the three repositories an accelerated growth curve in the most recent period, between 2018 and 2020. Considering that the most widely disseminated work on DSR is that of Hevner et al. (2004), published in a journal of the information systems area (*MIS Quarterly*), we can interpret the distribution of the data in Table 4.2 as resulting from a natural process of maturation and spread of the DSR approach to other areas of science beyond the area of Information Systems. The fact to be highlighted is that this is an approach in fast

Table 4.2 Longitudinal distribution of publications of DSR articles

Repository	Number of DSR articles published per year									
	2011	2012	2013	2014	2015	2016	2017	2018	2019	2020
EBSCO	12	8	18	17	20	22	24	31	46	79
ProQuest	1	4	3	1	4	3	8	6	11	20
PubMed	0	0	2	4	2	2	4	9	9	14
Total	13	12	23	22	26	27	36	46	66	113

pace of adoption by researchers from different areas, and should consolidate itself as one of the main approaches of pragmatic research during this third decade of the twenty-first century.

As an example of the application of the DSR approach in different areas, we describe below twelve artifacts developed for four different areas of science recognized as professional schools: administration, medicine, education, and engineering.

Examples of DSR Artifacts in the Field of Medicine

In Medicine we can think of an infinite number of artifacts, from physical ones, such as the design of equipment and physical tools for performing surgical intervention on bodies, to procedures (methods) for treatment. In related areas, such as pharmacy, we have the development of pharmaceuticals. One of the most notorious examples of innovation by exaptation is the drug Viagra, as already explained in the second chapter (Dew & Sarasvathy, 2016). Despite the many possibilities, the articles describing DSR artifacts for the health area are mostly related to the development of information systems in support of patients, managers, and health professionals. They are articles from professionals in the area of information systems who research and/or work in organizations in the health area. We present below three examples of recent articles of DSR approach that were conducted and published within the scope of the area and journals of the medicine:

a. Sampa et al. (2020) redesign the Portable Health Clinic (PHC), an Remote Healthcare Services (RHS), for the containment of the spread of COVID-19 as well as proposed corona logic (C-Logic) for the main symptoms of COVID-19;

b. Comesaña-Campos et al. (2020) developed a method that allows the early detection and prevention of potential hypoxemic clinical cases in patients vulnerable to respiratory diseases;

c. Lapão et al. (2017) presents the design of the ePharmacare artifact, an information system that allows pharmacists to provide online pharmaceutical care services. The artifact allows pharmacists to perform follow-ups on patients that regularly visit their pharmacies.

EXAMPLES OF DSR ARTIFACTS
IN THE FIELD OF EDUCATION

In the area of Education, the possibilities of application of the DSR approach are also numerous, just think about the intersection of the multiple dimensions of the context of education: (a) age range, from school-age children to the late education of the elderly; (b) people with disabilities, such as visual, hearing, mental, among others; (c) characteristics of the language used for education, mother tongue or second language; (d) characteristics of the student's environment, at school, at home, on the move; among many other contexts of interest when we consider different communities. Despite this broad set of opportunities, most DSR articles that mention the term education are directed to the business context, focused on the learning of employees, suppliers and customers.

We describe below three articles that effectively focus on Education in its most widespread sense, that of early childhood education, as well as groups less assisted by society. They are interesting examples of application of the DSR approach in these contexts, which despite being broad, is still little explored. The description of the three artifacts follows:

a. Laato et al. (2017) developed an application for teaching mathematics to children in primary schools from the composition of musical rhythms. Thus, the artifact collaborates with the teaching of the relationships between mathematics and music, allowing users to compose their own songs using numbers;

b. Hedvall and Svensson (2017, p. 1) using gamification technology developed an app to address health and nutrition issues with pregnant girls in Ethiopia, one of the countries with the highest maternal and child mortality rates in the world;

c. Usener et al. (2012) developed a method-type artifact for the evaluation of students attending an undergraduate course in Computer Science, via Internet (E-assessment) with exercises that require high-level cognitive skills.

Examples of DSR Artifacts
in the Area of Administration

As van Aken and Romme (2009) stated, organizations are also inventions of the human mind, they are artificial entities and can be considered human artifacts. They are composed of a multitude of artifacts, either to carry out their operations or for the activities of monitoring, evaluation and evolution of the work. In 2009, when the DSR approach began to be disseminated to other areas, van Aken and Romme (2009, p. 10) published an article in a journal of the Administration area highlighting three examples of artifacts directed to the area:

> Rozemeijer (2000), developing solution **concepts for organizing the purchasing** function of an industrial company; Andriessen (2003), developing solution **concepts for the valuation of the intangible assets of** an organization and Romme and Endenburg (2006), developing solution concepts for **re-engineering organizations** around the notion of circularity.

As we could observe in the data in Table 4.1, currently, journals from the Administration area are in second place in the ranking of those that most publish articles developed with the DSR approach, second only to the area of Information Systems. We present below three examples of recently published DSR artifacts within the context of the Administration area:

a. Afflerbach et al. (2016) developed a decision model that provides guidance on how to determine an economically appropriate business process standardization level for a business process;
b. Bemelmans et al. (2013) developed a model for maturity analysis of the purchasing process in organizations organized by projects, especially for organizations that operate in the construction sector;
c. Dos Santos and Da Silva (2015) developed a pricing model capable of helping companies providing services in information technology to offer flexible outsourcing contracts as to how to charge for their services.

Examples of DSR Artifacts
in the Area of Engineering

When thinking about all the diversity of engineering sub-areas (naval, forestry, aeronautical, mining, nuclear, automotive, chemical, mechanical, among others) we could glimpse several types of artifacts, including physical. When examining the DSR artifacts published by the journals of the engineering area we observe that there is a very large predominance of artifacts in the form of logical abstractions, with most of them focused on the area of software engineering. This is consistent and corroborating with our perception of previous research. Of the DSR articles collected from the research conducted in 2018, described in the subsection "Evolutionary cycles of DSR according to the different types of artifacts" (Chapter 3), we identified only one article that presented an artifact of physical nature: a "laptop trolley" used in a hospital environment to facilitate health professionals to access patient data (Weeding & Dawson, 2012). One of the possible explanations is the large concentration of the practice of the DSR approach around professionals in the area of information systems.

We present below three examples of DSR artifacts recently published in specific journals of the Engineering area and, evidently, of intangible nature in the form of logical abstractions:

a. Brady et al. (2018) developed visual tools, including physical displays, to discuss the model of production planning and control known as Lean construction management (LCM) model. These visual tools allow improving the flow of information, thus improving transparency between the interfaces of planning, execution and control;

b. Bjarnason et al. (2019) developed the Gap Finder method that is directed to increase requirements-test alignment in software development projects, that is, to make the specifications of the requirement engineering are contemplated in the testing phase;

c. Giménez et al. (2020) developed the Value Analysis Model (VAM) to measure the value creation expected by customers and to identify the value losses in the design process. Thus, the VAM allows the measurement and analysis of value through desired, potential, and generated value indexes, value loss identification, and percentages of value fulfillment in the design stage.

REFERENCES

Afflerbach, P., Bolsinger, M., & Röglinger, M. (2016). An economic decision model for determining the appropriate level of business process standardization. *Business Research, 9*(2), 335–375.

Bemelmans, J., Voordijk, H., & Vos, B. (2013). Designing a tool for an effective assessment of purchasing maturity in construction. *Benchmarking, 20*(3), 342–361.

Bjarnason, E., Sharp, H., & Regnell, B. (2019). Improving requirements-test alignment by prescribing practices that mitigate communication gaps. *Empirical Software Engineering, 24*, 2364–2409.

Brady, D. A., Tzortzopoulos, P., Rooke, J., Carlos, T. F., & Tezel, A. (2018). Improving transparency in construction management: A visual planning and control model. *Engineering, Construction and Architectural Management, 25*(10), 1277–1297.

Comesaña-Campos, A., Casal-Guisande, M., Cerqueiro-Pequeño, J., & Bouza-Rodríguez, J.-B. (2020). A methodology based on expert systems for the early detection and prevention of hypoxemic clinical cases. *International Journal of Environmental Research and Public Health, 17*(22), 8644.

Dew, N., & Sarasvathy, S. D. (2016). Exaptation and Niche construction: Behavioral insights for an evolutionary theory. *Industrial and Corporate Change, 25*(1), 167–179.

Dos Santos, J. C., & Da Silva, M. M. (2015). Price management in IT outsourcing contracts. The path to flexibility. *Journal of Revenue and Pricing Management, 14*(5), 342–364.

Giménez, Z., Mourgues, C., Alarcón, L. F., Mesa, H., & Pellicer, E. (2020). Value analysis model to support the building design process. *Sustainability, 12*(10), 4224.

Hedvall, A., & Svensson, E. (2017). *Teaching maternal healthcare and nutrition in rural Ethiopia through a serious game* (Dissertation, Malmö högskola/Teknik och samhälle). Retrieved from http://urn.kb.se/resolve?urn=urn:nbn:se:mau:diva-20939

Hevner, A. R., et al. (2004). Design science in information systems research. *MIS Quarterly, 28*, 75–105.

Laato, S., Laine, T. H., Seo, J., Ko, W., & Sutinen, E. (2017). Designing a game for learning math by composing: A Finnish primary school case. *2017 IEEE 17th International Conference on Advanced Learning Technologies (ICALT)*, pp. 136–138.

Lapão, L. V., da Silva, M. M., & Gregório, J. (2017). Implementing an online pharmaceutical service using design science research. *BMC Medical Informatics and Decision Making, 17*(1), 1–14.

Sampa, M. B., Hoque, M. R., Islam, R., Nishikitani, M., Nakashima, N., Yokota, F., Kikuchi, K., Rahman, M. M., Shah, F., & Ahmed, A. (2020). Redesigning portable health clinic platform as a remote healthcare system to tackle COVID-19 pandemic situation in unreached communities. *International Journal of Environmental Research and Public Health, 17*(13), 4709.

Usener, C. A., Majchrzak, T. A., & Kuchen, H. (2012). E-assessment and software testing. *Interactive Technology and Smart Education, 9*(1), 46–56.

van Aken, J. E., & Romme, G. (2009). Reinventing the future: Adding design science to the repertoire of organization and management studies. *Organization Management Journal, 6*(1), 2–12.

Weeding, S., & Dawson, L. (2012). Laptops on trolleys: Lessons from a mobile-wireless hospital ward. *Journal of Medical Systems, 36*(6), 3933–3943.

CHAPTER 5

Design Science Research Method

There are many proposals for design science research method (DSRM), but the most widespread is that of Peffers et al. (2007). The most recent DSRM proposals are all based on the text by Peffers and his colleagues. Thus, to facilitate comparison and analogies of our understanding of DSRM, we structured the content of this chapter according to the six phases of DSRM proposed by Peffers et al. (2007): Problem identification and motivation; Define the objectives for a solution; Design and development [of the artifact]; Demonstration; Evaluation; and Communication.

PHASE 1—PROBLEM IDENTIFICATION AND MOTIVATION

In the second chapter, when discussing the innovative artifacts, we observed that there are three creative actions of interest to the DSR according to the perspective of technological and commercial dimensions: invention, exaptation and improvement (Danneels, 2002; Gregor & Hevner, 2013, 2014). The artifact that innovates both technologically and commercially, called invention, is something very rare to occur either in the context of organizations or publications in scientific literature. Among the many reasons for the invention's greater complexity and risk is its extremely innovative perspective of the artifact, usually tied to a proposition to explore an opportunity perceived by the researchers. As noted earlier in this book, those involved in academic activities, such

© The Author(s), under exclusive license to Springer Nature Switzerland AG 2021
J. O. De Sordi, *Design Science Research Methodology*,
https://doi.org/10.1007/978-3-030-82156-2_5

as editors, reviewers, and referees for research funding agencies, tend to understand and agree more readily with problems recognized by the academic community. Thus, it makes sense that Peffers et al. (2007) label the first section of the DSRM as problem identification. It is important to note that this label does not adequately contemplate the logical trail of artifacts obtained by the creation action of the invention type, which is predominantly based on opportunities conceived by insights. The challenge is characterized by the fact that extremely innovative artifacts are not always perceived and required by practitioners. Thus, in addition to the greater creative demand for inventive artifacts, there is also the greater challenge to the adequacy and contextualization of this creative action for the context of the DSRM approach.

Problem identification in the context of DSR requires insertion of the proponent team with the field, experience with the professionals who will use the artifact (the practitioners), as well as a good command of the problem with which they deal. This particular demand is very similar to that of other qualitative research methods that require strong insertion of the researcher with the field of research. Often the problem manifests itself naturally to researchers who act or have acted as practitioners at some point. In research carried out in the areas known as professional schools, it is very common to find researchers who are also trained in the area itself, i.e., in addition to being researchers, they are or have been practitioners in the area. Thus, the perception of a field problem is naturally revealed to the researcher through practical experience obtained by one or more of the following ways: being or having acted at some point in the field; by talking with their peers, by participating in operational projects in the field, fairs, and congresses; or even receiving the provision of services or products prepared by their peers, i.e., placing themselves in the position of client. The greater this proximity of the researcher with the field and the practitioners, the greater the probability of perception and natural manifestation of a field problem.

Here it is worth rescuing the epistemological discussions of the relationship of proximity between researcher and its object of study, usually characterized by the insider versus outsider research dyad. According to Dwyer and Buckle (2009, p. 58).

Insider research refers to when researchers conduct research with populations of which they are also members (Kanuha, 2000) so that the researcher shares an identity, language, and experiential base with the study participants. (Asselin, 2003)

As the researcher or some of the members of the research team are part of the group of practitioners, who experience and know the difficulties of the field that the artifact is intended to assist, brings a number of advantages for the conduct of research. The main one is to promote the relationship, dialogue, and acceptance of the research team by the practitioners. The experience in the area brings to the team the domain of the technical jargon of the area, the knowledge of the artifacts currently used, greater understanding of the problems faced by practitioners, knowledge of important data, and indicators of the area, among other facilities. This helps to combat and reduce the social stigma of practitioners in relation to researchers, that is, it avoids researchers being stigmatized as external people and totally distant from the reality and problems of practitioners. Hence the indication of insider research literature brings legitimacy to researchers before practitioners, facilitating communication and access to them and other resources available in the field that are important for research DSR.

The initial analysis of the first phase of the DSRM proposed by Peffers et al. (2007), focused around the first noun used in its title, i.e., the term Identification. To this end, we highlighted the importance of distinguishing the identification of an important problem from the action of proposing an important solution. The second part of the name of the first phase—Problem identification and motivation"—involves the term motivation. The action of motivating is widespread in the context of scientific research. All new scientific knowledge reported should, in its introduction, be duly motivated, that is, the researcher should present to the readers the value, the actuality and the relevance of the problem addressed by his research.

Ellis and Levy (2008, p. 28) elaborated a set of procedures, called "problem statement template" to explain to novice researchers how to motivate and describe a research-worthy problem. The template indicates that in the first section of the scientific article there should be a set of citations to meet three demands: 1st) "list three current, peer-reviewed references (scientific articles) that support the presence of that problem and briefly describe the nature of that support"; 2nd) "list three current, peer-reviewed references that support the impact of the problem that the research proposes addressing and briefly describe the nature of that support"; and 3rd) "list three current, peer-reviewed references that support the conceptual basis of the problem and briefly describe the nature of that support."

Motivation is a logic of transparency and valorization of textual communication that occurs in several areas, not only in the context of scientific academia. For example, in the area of Public Administration, the practices and laws in force require public managers to motivate their opinions in response to requests from citizens and entities. For example, the administrative acts of public managers in European Union (EU) countries are guided by general principles of good administration as recommended in the European Code of Good Administration. To adequately meet the requirements of transparency and responsibility in the public process, the European Code of Good Administration recommends its member countries to adopt the "Principle of Motivation," as highlighted by Negrut (2011, p. 10):

> The principle of motivation requires the need for the authority issuing an administrative act to show explicitly the facts and law elements that determine the adoption of that decision' (Tofan, 2006, p. 46). 'Motivation is an essential element for the formation of people's conviction on the legality and appropriateness of the administrative act, representing also a guarantee of having chosen the optimal solution by the decision-making body' (Tofan, 2006, p. 46).

The principle of motivation applied in the context of the DSRM involves bringing quotes that highlight that there is a current, important and representative problem for a group of practitioners. This implies that at the end of the work of the first phase of the DSRM we have to have an essay that clearly highlights a current and relevant problem for a group of practitioners. This problem is perceived as current, that is, not yet solved and not fully supported by the current artifacts. Thus, the generated DSRM text at the end of its first phase should inform the reader outside the research team very clearly: the problem faced by a practitioner group, the definition of the practitioner group, the core activities and deliverables of this group, the basic functionality of the artifact that will be useful to practitioners, the main constituent elements of the new artifact. All of this should be well substantiated through indicator numbers that are recognized as important to the practitioner group, evidencing that there is a problem (underperforming indicator). Thus, the presentation of a problem within the DSRM context should also contain citations and references to important documents from sources relevant to practitioners (gray literature).

PHASE 2—DEFINITION OF THE OBJECTIVES FOR A SOLUTION

Once a research problem or opportunity has been identified, the researcher must plan the development of the artifact. Peffers et al., (2007, p. 54) indicated that this second phase of the DSRM can be inspired by the question "what would a better artifact accomplish?" To this effect, the researcher must describe the differential aspects of his artifact: (a) a new function or sub-function, not yet addressed by current artifacts; and/or (b) improved operational, financial, or qualitative performance. In this aspect it is observed the need for an exploratory research of similar artifacts available for the same class of problem. For each artifact identified, researchers should analyze the differential aspect of the new artifact, either in terms of available function or performance. The results of this comparison are usually summarized at the beginning of the DSR article. Below the analogy developed for the artifact we are using as an example, the AnaCoTEx (De Sordi et al., 2016, p. 901):

> By comparing AnaCoTEx with the available textanalysis software (TAS), its distinctive and innovative aspects are highlighted: (a) The TAS analysis of cohesion is linear and considers adjacent sentences, whereas AnaCoTEx performs a cross-analysis between all elements of the text structure, that is, a square matrix including all chapters, subchapters, and other levels in the structure; (b) TAS analyzes text words in their syntactic and lexical aspects without considering the distribution of the text in terms of the structure of chapters and subchapters, whereas AnaCoTEx focuses exclusively on analyzing the text structure without addressing syntactic and lexical aspects; (c) TAS has restrictions in terms of the number of characters or words in the text to be analyzed (approximately 15,000-20,000 characters); with AnaCoTEx, utility is proportional to the size (i.e., the larger and more structured that a text is, the greater the earned value); (d) TAS is primarily designed for English language texts, whereas AnaCoTEx is multilingual; and (e) TAS does not require prework of the text to be analyzed; AnaCoTEx requires the insertion of cross-references in the form of numbers in brackets. In short, the features and functions of AnaCoTEx are distinct and focused specifically on the cohesion of extensive texts.

AnaCoTEx does not compete with TAS: On the contrary, it complements TAS with additional analyses.

It is important to think at this moment of planning in the characteristics to be delivered by the new artifact, hence the use of the term in plural in the title of this phase, explaining the idea of going beyond the general objective, covering the specific objectives. As addressed in the strategic planning works of companies or other research methods, each specific objective can be considered as an action or work front. Following the recommendations of planning methods, each characteristic or work front must be analyzed before two dimensions: the characteristic's importance to practitioners and feasibility in terms of being technically and financially feasible. Following these recommendations avoids the design of unfeasible parts, focusing on what is feasible, pointing out those few things that effectively must be fully met and delivered during the execution of the project. In a more extreme situation, it may even identify the unfeasibility of the development of the artifact itself. From this activity it is derived the indicators to be used for the evaluation phase of the artifact, i.e., once instanced in operation, there is an expectation of functions to be performed (qualitatively) and results to be achieved (quantitatively). Here there is also the adherence with the specific planning methods, which indicates the need for an indicator for each specific objective defined.

As noted in the Design Theory subsection of the first chapter, DSR produces mid-range theory (Van Aken & Romme, 2009). Design theory or mid-range theory is composed of several components, the higher level ones being labeled as "meta-requirements which describe the class of goals to which the theory applies" (Walls et al., 1992, p. 42). Meta-requirements are not always easy to be observed, considering that DSR articles usually publish only the description of the specification of the new artifact, the meta-design. This lack of information is often due to the limitations of the quantity of words of the scientific communications in the form of article. As an example, we have in Table 5.1 the meta-requirements and the meta-designs of a DSR project that developed an artifact of construct type, in the form of taxonomy, to discuss creative logics practiced by entrepreneurs and called "10 Types of Creative Reasoning." Thus, the first column of Table 5.1, can be understood as specific objectives to be achieved during the development and evaluation cycles of the artifact. The artifact generated by the research associated with Table 5.1 is available at https://tentypescreation.com/.

Table 5.1 Example of specific objectives in the form of meta-requirements

Meta-requirements (The artifact must…)	Meta-designs (For this, its design must…)	Contextualization (In the design process that was achieved)
Demonstrate that creation is possible through adaptation not only of the form but also the function of a product or service	Bring to the central display of the artifact a dichotomy centered on the subject of "alteration," with one of the options being the "form" and the other the "function"	Dichotomy of the "useful" subject, distinguishing the "exaptation" tactic from five others, located below the hierarchy of the subject of "alteration of the form"
Demonstrate that when working with renewal it is necessary to consider not only the things that are deemed useful but also less desirable things, viewed as useless by the organization	Bring to the central display of the artifact a dichotomy centered on the subject of "aptitude [of the existing resource]," with "useful" as one of the options and "useless" as the other"	Dichotomy between the "exaptation" tactic and the others below the hierarchy of the subject of "alteration of the form"
Demonstrate that the alteration of the form of a product or service may have different goals: to alter performance, reduce costs or to make the product or service more adaptable to the needs of the end-user	Bring to the central display of the artifact a trichotomy centered on the subject of "orientation [of the alteration to the form]," indicating the options "cost," "performance," and "customer"	Trichotomy that presents five creative tactics associated with alteration of the form: custom-made, adaptation, frugal, improvement, and degradation
Demonstrate that the alteration of performance of a product or service does not always mean an improvement; the opposite can also occur, with a reversal, a reduction	Bring to the central display of the artifact a dichotomy centered on the subject of the "direction [of the alteration of performance]," indicating the options of "superior" and "inferior"	Dichotomy between the "improvement" tactic and the "degradation" tactic
Observe that texts on creativity stem from different fields of science. Thus, it is necessary to be aware of the ontological challenges involved in describing terms and concepts from fields that are very different from those in which the practitioners are qualified	Include complete definitions with references (if further information is required), presenting meaningful examples, in other words, as close as possible to the reality of the practitioners	In the development of the artifact, we worked with many terms from the field of biology that required us to search for specific examples in the field of innovation

(continued)

Table 5.1 (continued)

Meta-requirements (The artifact must...)	Meta-designs (For this, its design must...)	Contextualization (In the design process that was achieved)
Use a content presentation structure that makes it easy to understand and that encourages the use of the artifact	Structure the central content of the diversity of the creative tactics using a figure that is easy to understand	We used a very simple and widely used taxonomic structure, a folk taxonomy (Miles & Huberman, 1994)
Facilitate the identification of the common and differential aspects between the diverse creative tactics that are addressed	Select a type of display that facilitates the perception of common inherited characteristics as well as divergences	We adapted the folk taxonomy with the inclusion of labels referring to the differentiating subjects, seen in the upper part of each dichotomy and trichotomy in the taxonomic structure
Encourage and facilitate the use of the artifact, making the first actions to understand the informational artifact include content of which the practitioners have greater knowledge and mastery	Prioritize the Copy creative tactic as a means of explaining the others, as it is the most widely diffused among the practitioners and end-users	We began our taxonomic structure with the Copy creative tactic placed in the highest part of the structure, the first part to be read
Add functionalities to the artifact that enable not only a better understanding and mastery of the creative tactics, but also greater utility, perceived value and willingness to use on the part of the practitioners	Incorporate into the artifact content on the creative tensions helping to explain the framework of the taxonomic structure (the dichotomies and trichotomies). This highlights the importance of the creative tactics and allows the practitioners to explore themes that are important to their activities along with the end-users	We included content in the artifact regarding the creative tensions: effectuation versus causation, and exploitation versus exploration
Demonstrate and address adequately the complexities and variations of each creative tactic	Explain the core actions where there is more than one implementation strategy for a creative tactic	In addition to clarifying the definition of the tactic, we presented an example for each action. This occurred in the case of three creative tactics: degradation, frugal, and nonaptation spandrels

(continued)

Table 5.1 (continued)

Meta-requirements (The artifact must...)	Meta-designs (For this, its design must...)	Contextualization (In the design process that was achieved)
Work the definitions of how necessary it is to avoid misunderstandings regarding the content of the articles by the practitioners	Use contrast tables (or similar resources) that demonstrate the similarities and differences between the creative tactics that are causing misunderstandings on the part of the practitioners	We created a contrast table for two creative tactics: Degradation and Frugal
Show that there is a great diversity of creative tactics besides the NPD tactic, ones that are more in keeping with the reality of small companies	Present in the informational structure of the taxonomic structure of the article as soon as possible (in the highest possible part of the structure) the difference between innovation and renewal, in other words, between radical innovation and innovation of what is already on hand	Soon after the dichotomy of insight by copy, we already present the dichotomy of the existing central resource, in other words, from things that the entrepreneur or small business manager already possesses

Phase 3—Design and Development

The purpose of this phase is to create the artifact from the functionalities (meta-requirements) defined in the previous phase, arising from the specific objectives declared for the artifact. Among the projects developed by professionals in the various areas of science, a very widespread and familiar type of design is the architecture for buildings. Among the objectives for the housing of a family, it is possible to imagine the adequacy for a dweller with mobility difficulties that would imply in meta-designs, for example, of a terraced building with the presence of a lift or a ground floor building. From the initial objectives (meta-requirements), the architect develops and presents successive versions of house projects (meta-designs) for analysis and discussion with his clients. Thus, the house design is being adjusted from this interaction between architect and clients. This interactive process occurs in this third phase of DSRM, called by Hevner et al. (2004) as generate/test cycle. To meet a meta-requirement the designer identifies and tests several meta-design options, hence the use of the term *testable design product hypotheses* "which can be used to verify whether the meta-design satisfies the meta-requirements" (Walls et al., 1992, p. 43). This interactive process with generation of versions of the design is repeated until the moment in which the designer understands to have reached the functional saturation for all the meta-designs of the artifact. At this moment there is the understanding that the characteristics present in the project are sufficient to solve the problem in a full and comprehensive way, able to meet that class of problem.

DSRM is recognized as a pragmatic or problem-solving research paradigm (Hevner et al., 2004) whose central questions and purpose have been well summarized by Goldkuhl (2012, p. 92): "questions of utility and effectiveness are in pragmatism always accompanied with questions of appropriateness of ends." Thus, the emphasis of the generate/test cycle is on ascertaining and ensuring appropriateness of ends. Returning to the example of the artifact "10 Types of Creative Reasoning" (Table 5.1), several questions were asked by the development team during the generate/test cycle aiming to increase the utility and effectiveness of the artifact. As the users of the artifact are the consultants and lecturers (in the role of practitioners or users of the artifact) who work together with entrepreneurs and future entrepreneurs (in the role of clients or final beneficiaries of the artifact), the suitability of the artifact as an instrument of interaction between these two groups of actors was very much observed. Since the proposed artifact is a tool in the form of a taxonomy, aimed at the dialogue between practitioners and customers about possibilities of creative logics for diversification of the family of products and services of companies, several specific objectives (meta-requirements) were defined thinking of functions that facilitate the communication between the two groups of actors involved.

One of the examples of meta-requirements linked to specific communication objectives between the actors is the "encourage and facilitate the use of the artifact, making the first actions to understand the informational artifact include content of which the practitioners have greater knowledge and mastery" (stated in the eighth line of Table 5.1). The insight for the definition of this meta-requirement was the good practices for the elaboration of textual communication aiming to facilitate the understanding of the reader. One of these very common practices is to "begin sentences with familiar (old) information and conclude sentences with unfamiliar (new) information" (Duke, 2013, p.1). From the meta-requirement in question, it was observed necessary characteristics to the design of the artifact or meta-designs, among them: "prioritize the Copy creative tactic as a means of explaining the others, as it is the most widely diffused among the practitioners and end-users." The definition of this meta-design, concerning how to initiate the structuring of the taxonomic tree was ground on the group's knowledge that among the ten creative logics defined, exemplified, and compared in the artifact, Copy is the simplest and of low cost (Ofek & Turut, 2008). Consequently, Copying presents itself as the most well-known and simplest creative logic to understand. In the eighth row of Table 5.1, in the contextualization column, it is stated how this meta-design was implemented in the artifact: "we began our taxonomic structure with the Copy creative tactic placed in the highest part of the structure, the first part to be read."

Obviously this detailed breakdown between meta-requirements, meta-designs, and contextualization of the artifact, as stated in Table 5.1, is not present in DSRM articles. This is an effort in unveiling the sequencing of concepts and entities important to DSRM within the logic employed by researchers for the development of their artifacts. In a more simplistic description they can be understood as responses to the sequencing of the following expressions: "the artifact must...," for the meta-requirements of the artifact; "for this, its design must...," for the meta-designs of the artifact; and "in the design process that was achieved ...," for the contextualization of what the design team accomplished or will do. The central objective here is to observe that the meta-requirements generate meta-designs and that these demand actions of contextualization of the artifact design.

Although the meta-designs presented in Table 5.1 are textual, there are different ways of expressing them. Depending on the communication patterns employed by different groups of professionals, different forms may be adopted. Engineers use different techniques to express their concepts. In software engineering, we can express the meta-designs of an information system by the integrated development of different techniques such as: interaction diagram, entity-relationship model, state transition diagram, process decomposition diagram, algorithms, among others. When reading a paper developed with the DSR approach these meta-designs expressions are not always present or easily identified. The major ontological efforts of DSR article authors have been in describing the operational procedures for using the new artifact and its efficiency in terms of usefulness. The focus is prescriptive, on presenting solutions. However, with the meta-designs not made explicit or not very explicit, there is difficulty in advancing in terms of discussion and evolution of the artifact itself. It would be important that editors of journals of professional schools request the authors of research developed with the DSRM that they explicit in their articles the meta-designs of the artifact, even if in the form of supplementary material to the article.

Phase 4—Demonstration

Conceived the artifact design the next step is the use of the artifact. Here the goal is the integrated testing of all concepts and components of the artifact, the realization of a comprehensive proof-of-concept. The goal is to find out if the artifact is able to solve one or more instances of the problem category associated to it. Just as the algorithm of a software can be previously tested by "table test," without having a single line of source code written, the artifact can also be tested before it is even built. Those responsible for the artifact can test the artifact in several ways, Peffers et al. (2007, p. 55) highlighted: "experimentation, simulation, case study, proof, or other appropriate activity."

Among the aspects to be verified with the demonstration are the procedures for the use of the artifact, covering here the necessary inputs; the procedures performed by the practitioner, by the beneficiary of the artifact, and those performed by the artifact itself; the outputs or results provided; and some of the performance indicators possible at this time. Indicators of accuracy are often possible, but not of execution time, considering that it is not yet the final and ideal platform, only an environment for demonstration/testing of the artifact. The use of the artifact in this phase is something simpler, considering that its operation is done by the researchers themselves or well assisted by them. For the next phase of evaluation, there is the need for greater readiness of the artifact, because it should be used in its natural environment, by the typical practitioner and without any interaction with the researcher. The analysis of results achieved with the demonstration may indicate the need to return to the previous phase with the purpose of reviewing and changing the artifact meta-designs.

Although many authors indicate that the works performed in their research with DSRM are configured as artifact evaluation, most of the tests performed are configured more as an artifact demonstration action. Usually they are not fully completed versions of the artifact, they are artifacts in more embryonic stages that hinder the realization of a full evaluation as it should be. There is a wide range of aspects to be considered for an evaluation process. Thus, although few studies with application of DSRM declare the demonstration phase, opting for the term evaluation, we have that the demonstration is something much more common to occur than formally declared in the articles. In the next subsection we will describe the several aspects necessary for configuration of an evaluation process, evidencing the hyper declaration of the evaluation procedures, as well as the sub declaration of the demonstration actions.

For this confusion of naming actions performed some justifications can be explored. In the structural issue of writing reports and articles associated with DSR, Gregor and Hevner (2013) indicated seven titles of text sections as central elements of a "Publication Schema for a Design Science Research Study." The seven denominations presented are: introduction, literature review, method, artifact description, evaluation, discussion, and conclusions. It is observed that there is no section named demonstration. In extreme situations in terms of innovation, involving "very novel artifacts, a 'proof-of-concept' may be sufficient" (Gregor & Hevner, 2013, p. 351). In short, the evaluation section is not always mandatory, and may be replaced by a demonstration section of the artifact. Thus, many authors afraid of having their research labeled as partial, may have mistakenly chosen to use the term evaluation instead of the term demonstration for designation of the tests performed with the artifact.

PHASE 5—EVALUATION

Within the context of DSRM terms "more approachable and less confusing" (Baskerville et al., 2015, p. 560), there are works exclusively focused on the discussion of the deficiencies and concerns with the process of artifacts evaluation. In this sense, the research of Venable et al. (2016, p. 77) expresses this concern well, by stating that "the extant DSR literature provides insufficient guidance on evaluation to enable Design Science Researchers to effectively design and incorporate evaluation activities into a DSR project that can achieve DSR goals and objectives." Venable et al. (2016) defined the dimension "paradigm of the evaluation study" in which the extreme options are: naturalistic evaluation and artificial evaluation, with only the former adhering to DSR principles. The shortcomings of artificial evaluation are well described in Venable et al. (2016, p. 81):

artificial evaluation involves reductionist abstraction from the natural setting (in order to assure rigour in its assessment of efficacy of the Technology artefact) and is necessarily unrealistic in the sense that it fails to adhere to one or more of the three realities (i.e., unreal users, unreal systems, or unreal problems) of Sun and Kantor (2006).

Of the three realities necessary for the evaluation process—"real users, real problems, and real systems" (Sun & Kantor, 2006, p. 616)—the perspectives of real users and real systems can be structured and analyzed within a user-centered perspective. Considering the criticism of artificial evaluation and the demands of a naturalistic evaluation we identified four central aspects for an evaluation process, according to the assumptions of the DSR, which are described in Table 5.2. For naturalistic evaluation there is an effort by researchers to deliver the artifact to practitioners which, obviously, involves some level of basic instruction for operating the artifact. Depending on the differences between the proposed artifact and current artifacts there may be a need for training of practitioners to enable them to operate the new artifact.

An analysis of the assumptions of DSRM that characterize it as a pragmatic research approach, with the development of artifacts aimed at utility and effectiveness for professionals facing specific problems, implies full utilization, as occurs in real life. Thus, we have this fifth phase of DSRM, the evaluation, as vital to the success of the DSR strategy itself. Considering the relevance of this theme, we will return to it in the sixth chapter, where we will address the current stage of the evaluation process of the DSR artifacts as stated in the recently published articles.

Table 5.2 Central aspects for evaluation of an artifact according to DSRM

	Central Aspects	*Guiding question for analyzing the evaluation of the artifact*
Artifact Evaluation	Use of the artifact	Was the artifact effectively used in its core function?
	User tester	Was the artifact operated by a typical user?
	Environment of use	Was the artifact used in an appropriate environment of typical users?
	Inputs for use	Were data or other resources of the appropriate environment used for the tests?

Once the naturalistic evaluation of the artifact is assured, this phase ends with the analysis of the results achieved by the use of the artifact in the field. Here are used the indicators defined in the second phase, "definition of the objectives for a solution." Thus, as in the previous phase of demonstration, the analysis of these indicators of the artifact may result in changes and in new propositions of meta-designs for the artifact, i.e., the return to the activities of the third phase, "design and development."

Phase 6—Communication

The dissemination of DSR should take into account two types of communication for two groups of readers, one consisting of practitioners or users of the proposed artifact, and the other composed of researchers of the professional school associated with the artifact and the group of practitioners. Thus, researchers must deal with two sets of content, for publication in distinct outlets, demanding communication patterns quite specific to each. The communication style for practitioners is quite different from scientific writing, and can be considered as a challenge to researchers practicing the DSR strategy. Even the text of scientific communication, which is closer and more familiar to researchers, presents several peculiarities for the context of DSR, making it equally challenging. Thus, there are two communication challenges for those who practice DSR, both of which will be explored in this subsection.

Communication with practitioners is not demanded for the vast majority of research strategies. For the DSR it is fundamental in order to ensure one of its greatest benefits, that of accelerating the process of transition from invention to innovation. The effective use of the artifact generated by the DSR must occur in the field, by the practitioners and, for this, the practitioners must become aware of the existence of the artifact. Thus, identifying the preferred media of the artifact's practitioner audience is essential to realizing the full potential of DSR. Texts for practitioners should pay more attention to pragmatic issues, such as: purpose and advantages of using the artifact, highlighting the problems solved by the artifact; the way to access the artifact or, eventually, the instructions on how to reproduce the artifact; and mainly the instructions on how to operate and make use of the artifact. As highlighted in the first chapter, in Fig. 1.1, this type of literature for practitioners has a greater focus on practicality, on the applicability of the artifact.

When discussing the DSR applied to the area of Administration, van Aken (2005, p.19) highlighted that communication with practitioners can help researchers deal with an old and thorny difficulty of the area: "the relevance problem of academic management research in organization and management." Thus, we have that the communication strategy of research DSR covering practitioners can be perceived as an instrument for strengthening and legitimizing areas of science, particularly those associated with professional schools.

For communication directed to the scientific community there are some very specific demands of the DSR approach in relation to the format used in traditional positivist or hypothetical deductive research. Traditional research is usually structured in six parts: (1) Introduction, (2) Literature review, (3) Methodology, (4) Result, (5) Discussion, and (6) Conclusion (Sun & Linton, 2014). The DSR strategy requires that the communication of the newly generated knowledge includes two very specific sections: "artifact description" and "evaluation/demonstration." We describe below the contents of these two sections of the scientific report or article associated with DSR.

Regarding the content of the "artifact description" section Gregor and Hevner (2013, p.350) highlighted:

> Several sections may be needed and will likely occupy a major portion of the paper. The format may be variable but should include at least the description of the design artifact and, perhaps, the design search (development) process that led to the discovery of the artifact design. Presenting the design search process might assist in demonstrating credibility.

It is important to note the optionality characterized by the text passage "perhaps, the design search (development) process." This uncertain situation can be understood due to different levels of complexity of the DSR project, resulting from the proposition of one or more artifacts of different types (construct, model, method, and instantiation), covering different levels of domain and disclosure through previous articles. Thus it is interesting to observe the possibility of the description of these two work fronts that demand ontological efforts of DSR: the description of the artifact itself, and the description of its development process. These two aspects are well characterized in the text of Walls et al. (1992) that pointed out two components necessary for the description of knowledge generated by DSR, the design product and the design process. The design product comprises four contents: meta-requirements, meta-design, kernel theories (governing design requirements), and testable design product hypotheses. The design process covers three contents: design method, kernel theories (governing design process itself), and testable design process hypotheses.

Regarding the content of the other specific section of the DSR indicated by Gregor and Hevner (2013), the "evaluation," we have that it can be circumstantially replaced by the Demonstration section, as already discussed at the end of the subsection "Phase 4 – Demonstration." It is understood from the discussions of this subsection that having occurred the artifact evaluation, the content related to the "testable design product hypotheses" is mandatory. This implies describing the procedures used to put in use and test the artifact, as well as the results achieved from its use. It is important to compare the results achieved by the use of the new artifact, with the results of the other artifacts in use so far. This comparison should make use of one or more indicators of wide domain and use by practitioners, this simplifies communication with practitioners and makes it simpler to evidence the usefulness of the artifact. If the tests are only in terms of demonstration, there is no discussion of indicators, but of the findings from the broad proof-of-concept.

DSRM Summary

Table 5.3 presents a summary of the main activities and products generated throughout the six phases of the DSRM. It is important to remember that there is a strong interaction between the phases in both directions, i.e., each new phase may identify needs of researchers to return to activities of previous phases, complementing or even changing their outputs. The knowledge generated from each phase is cumulative, serving as input for the following phases. Thus, we describe in the input column only the inputs resulting from the initial preparatory work of that phase, and all the outputs of previous phases should be added to these inputs.

Table 5.3 Key inputs and outputs of the six stages of DSRM

Input	Phase	Output
Experience of researchers with the practitioners (insider); Domain of the problem	1. Problem identification and motivation	Definition of the problem, the practitioner group, the core activities and deliverables of that group, the basic functionality of the artifact that will be useful to practitioners, the main constituent elements of the new artifact, and indicators that characterize the problem
Limitations of current artifacts	2. Define the objectives for a solution	General objective; Specific objectives; and Meta-requirements
Meta-requirements	3. Design and development	Meta-designs
Logical and/or physical construction of the artifact from the meta-requirements	4. Demonstration	Artifact analyzed and "tested" from the actions of researchers
Operational artifact; Instruction/training of practitioners	5. Evaluation	Analysis of the performance of the artifact according to the values of indicators calculated from data obtained by the use of the artifact in the field by practitioners
How to use; expected and achieved results; problem, objective, and motivation; design product (meta-requirements, meta-design, kernel theories governing design requirements, and testable design product hypotheses); and design process (design method, kernel theories governing design process itself, and testable design process hypotheses)	6. Communication	Texts for outlets aimed at practitioners; article for outlets aimed at researchers

REFERENCES

Asselin, M. E. (2003). Insider research: Issues to consider when doing qualitative research in your own setting. *Journal for Nurses in Staff Development, 19*(2), 99–103.

Baskerville, R. L., Kaul, M., & Storey, V. C. (2015). Genres of inquiry in design-science research: Justification and evaluation of knowledge production. *MIS Quarterly, 39*(3), 541-A9.

Danneels, E. (2002). The dynamics of product innovation and firm competencies. *Strategic Management Journal, 23*(12), 1095–1121.

De Sordi, J. O., Meireles, M., & de Oliveira, O. L. (2016). The Text Matrix as a tool to increase the cohesion of extensive texts. *Journal of the Association for Information Science and Technology, 67*(4), 900–914.

Duke University. (2013). *Graduate school, scientific writing resources.* Retrieved from https://cgi.duke.edu/web/sciwriting/index.php?action=lesson2#principles

Dwyer, S. C., & Buckle, J. L. (2009). The space between: On being an insider-outsider in qualitative research. *International Journal of Qualitative Methods, 8*(1), 54–63.

Ellis, T. J., & Levy, Y. (2008). Framework of problem-based research: A guide for novice researchers on the development of a research-worthy problem. *Informing Science, 11*, 17–33.

Goldkuhl, G. (2012). Design research in search for a paradigm: Pragmatism is the answer. In M. Helfert & B. Donnellan (Eds.), *Practical aspects of design science* (Vol. 286, pp. 84–95). Springer.

Gregor, S., & Hevner, A. (2013). Positioning and presenting design science research for maximum impact. *MIS Quarterly, 37*(2), 337–355.

Gregor, S., & Hevner, A. (2014). The Knowledge Innovation Matrix (KIM): A clarifying lens for innovation. *Informing Science: THe International Journal of an Emerging Transdiscipline, 17*, 217–239.

Hevner, A. R., March, S. T., Park, J., & Ram, S. (2004). Design science in information systems research. *MIS Quarterly, 28*, 75–105.

Kanuha, V. K. (2000). "Being" native versus "going native": Conducting social work research as an insider. *Social Work, 45*(5), 439–447.

Miles, M. B., & Huberman, A. M. (1994). *Qualitative data analysis: An expanded sourcebook* (2nd ed.). Sage Publications.

Negrut, V. (2011). The Europeanization of public administration through the general principles of good administration. *Acta Universitatis Danubius. Juridica, 7*(2), 1–15.

Ofek, E., & Turut, O. (2008). To innovate or imitate? Entry strategy and the role of market research. *Journal of Marketing Research, 45*(5), 575–592.

Peffers, K., Tuunanen, T., Rothenberger, M. A., & Chatterjee, S. (2007). A design science research methodology for Information Systems Research. *Journal of Management Information Systems, 24*(3), 45–77.

Sun, H., & Linton, J. D. (2014). Structuring papers for success: Making your paper more like a high impact publication than a desk reject. *Technovation, 34*(10), 571–573.

Sun, Y., & Kantor, P. B. (2006). Cross-evaluation: A new model for information system evaluation. *Journal of the American Society for Information Science and Technology, 57*(5), 614–628.

Tofan, A. D. (2006). *Instituții administrative europene/European administrative institutions.* Bucharest (RO): C.H. Beck.

van Aken, J. E. (2005). Management research as a design science: Articulating the research products of mode 2 knowledge production in management. British *Journal of Management, 16*, 19–36.

van Aken, J. E., & Romme, G. (2009). Reinventing the future: Adding design science to the repertoire of organization and management studies. *Organization Management Journal, 6*(1), 2–12.

Venable, J., Pries-heje, J., & Baskerville, R. (2016). FEDS: A framework for evaluation in design science research. *European Journal of Information Systems, 25*(1), 77–89.

Walls, J. G., Widmeyer, G. R., & El Sawy, O. A. (1992). Building an Information System Design theory for vigilant EIS. *Information Systems Research, 3*(1), 36–59.

CHAPTER 6

Theory Development from Artifacts

This chapter explores the contributions of DSR projects beyond the development of artifacts (idiographic science), directed to the analysis of projects that resulted in the development of theories (nomothetic science). Aiming to better understand the conditions for the occurrence of this phenomenon, the content analysis technique was applied to 92 articles that proposed DSR artifacts and, subsequently, to another 2,389 articles that cited the former.

DSR Contribution Types

The use of a new artifact developed according to the principles of DSR "embodies design ideas and theories yet to be articulated, formalized, and fully understood" (Gregor & Hevner, 2013, p. 341). In other words, it has a potential for the continuity of research, including in the theoretical field. Baskerville et al. (2015) consolidated the perception of three groups of researchers evidencing three different levels of contribution from studies using a DSRM. Level 1, the most elementary and commonplace, is characterized by the proposition and instantiation of an artifact. At the third and highest level of contribution (Level 3), the design theory that encompasses the development of mid-range and grand theories is achieved. Gregor and Hevner (2013, p 346) described this stage as a "well-developed design theory about embedded phenomena," the

© The Author(s), under exclusive license to Springer Nature Switzerland AG 2021
J. O. De Sordi, *Design Science Research Methodology*,
https://doi.org/10.1007/978-3-030-82156-2_6

main result of which is "improved understandings of the problem and solution spaces" about the phenomena under study. Thus, the levels of contribution of DSR can range from a minor scientific contribution of less developed knowledge (Level 1) to a more abstract contribution of more elaborate and deeper knowledge (Level 3). The intermediate contributions (Level 2) will be described later.

When analyzing the potential of the DSRM for the generation of knowledge, Baskerville et al. (2015, p. 560) found that "much of the current empirical work in design-science actually engages science in an idiographic way, often involving some form of prototyping." It is important to observe the use of the term "prototyping," characterizing here an initial test stage of an artifact, in other words, before its effective use in the field by users understood as typical practitioners. Baskerville et al. (2015) characterized published DSR studies as predominantly at the most elementary level (Level 1), focusing on the presentation and instantiation of artifacts. They highlighted this situation as conflicting with researchers' preference and predisposition for nomothetic science (Level 3) and the lack of acknowledgment of idiographic science. This incoherent situation was identified by Baskerville et al. (2015) as a gap, a possibility for the continuity of studies involving the DSRM. Aware of the relevance of this gap due to the potential to generate scientific advances (Level 3) from already developed knowledge (Levels 1 and 2), we recognized and characterized it as an opportunity to make a scientific contribution.

Although the foundations and concepts of DSR date back to the nineteen sixties, the understanding and application of DSR as a mainstream research paradigm is a recent phenomenon. Peffers et al. (2018) traced its origin as a paradigm back to 2006, with the emergence of well design-centered conferences specifically directed toward the theme of DSR. As it has been only fifteen years since the publication of articles associated with studies using the DSRM began, and since six years have passed since Baskerville et al. (2015) referred to the lack of knowledge associated with theorization based on the insertion and use of artifacts, in this chapter I opted to analyze the levels of contribution of DSR studies with regard to theorization (Level 3). Thus, I intended to verify whether the gap identified by Baskerville et al. (2015) still persists. If I identify occurrences of theoretical scientific knowledge (Level 3) stemming from the analysis of data based on the use of DSR artifacts, I will explore the conditions in which this phenomenon occurred with the objective of gaining a better characterization and understanding of it.

Assumptions of DSRM

One of the premises of the DSRM is that the diverse contributions that could possibly occur in its projects do so over time, in phases, which are distributed into different articles. Baskerville et al. (2015, p. 542) claim that the knowledge generated by DSR is polymorphic and produced at different moments or in different episodes: "understanding knowledge production in design-science requires more than just the distinction of 'know-how' from 'know-why' (Kogut & Zander, 1997); in fact, episodes of both are necessary to justify knowledge production." They emphasize that the drafting of diverse manuscripts is the most appropriate way for us to have an adequate process of justification and evaluation for each knowledge moment associated with extensive and complex DSR projects. Thus, DSR projects are evolutionary, made up of diverse articles, with the knowledge moment associated with theoretical development (Level 3), which delivers more mature and developed scientific knowledge, being the last stage of the project to be achieved (Gregor & Hevner, 2013). This premise is the reason why, in this study, I conduct a longitudinal survey and analysis of articles associated with DSR projects, as I will address in the method section.

Kuechler and Vaishnavi (2008) conducted studies on theorization based on DSR projects. They claimed that for actions to refine the kernel of theories, involving an explanatory statement based on cause–effect relationships, it is fundamental to collect evidence from the effective use of the artifact, in other words, from the artifact's evaluation process. Thus, a second premise of DSR is that the reaction of end users in response to the use of the artifact is fundamental to achieve its theorization potential. To explore the reaction of users, some research strategies are more pertinent, such as the application of the grounded theory approach. Corbin and Strauss (1990, p. 5) highlight that "grounded theory seeks not only to uncover relevant conditions, but also to determine how the actors respond to changing conditions and to the consequences of their actions." The similarities between the DSR and grounded theory approaches are manifested even in the common reasoning between the scientific knowledge that they produce. Both design theory (Gregor & Hevner, 2013) and substantive theory (Glaser & Strauss, 1967) are founded on the construct of the Middle-range Theory of Merton (1949). Associating Level 3 of DSR with the substantive theory of grounded theory, it is possible to glimpse a wide range of possibilities for discussing the application of artifacts, involving at least two groups of actors: (a) practitioners/professionals, in other words, those who use or operate the artifact; and (b) clients/beneficiaries, in other words, those who receive the benefits of using the artifact.

The aspects addressed in this second premise show evidence that the effective use of the artifact by practitioners is fundamental to enable theorization based on cause–effect analyses within research projects using a DSRM. Therefore, we observe the importance of the means of analyzing and differentiating the effective use of a DSR artifact in other more simplistic ways of testing the artifact. Prototyping of the artifact by its developer and experimentation of the artifact by students are examples of characteristic actions in the phase of demonstrating the artifact, which precedes the evaluation phase that requires it to be used by practitioners (Peffers et al., 2007). To address the effective use of the artifact by practitioners, we used three analysis codes associated with three characteristics: user tester, use environment, and inputs for use (these codes have already been described in Table 5.2).

The third premise explored in this study addresses the criteria for identifying theorization or the presence of nomothetic knowledge in DSR projects. Theories are statements that can be applied to a class or kind of phenomena. Therefore, they must have the typical characteristics of nomothetic knowledge: applicability, generalizability, external validity, transferability, consistency, reliability, and dependability (Guba, 1981). It is recognized today that nomothetic knowledge is not necessarily associated with large samples and group-based statistics such as means and variances, and it can even be obtained using smaller samples, and a case-by-case form of analysis (Robinson, 2011). Thus, the most important characteristic for identifying nomothetic knowledge in the context of DSR projects is not the sample size or use of certain analysis techniques, but the identification of evidence that characterizes the generation or refinement of theories. This premise governed the actions of this study to gauge the presence of texts, in diverse articles associated with DSR projects that evidence a proposal for the refinement of theory or even the proposition of a new theory. We will not go into the question of whether the theory is well founded or formulated. We will concentrate on a stage prior to this, the intention of the authors of DSR projects to formulate and present theorizations based on data obtained through the use of the DSR artifact.

METHOD

Selection of DSR Articles to Compose the Sample

The articles associated with DSR were selected following a query in the Web of Science (WoS) repository. The WoS was chosen because it "represents a reasonable cross section of present science by indexing the most journals considered 'high quality', and that it serves equally well to highlight the most visible pieces of literature as any other individual database" (Piirainen et al., 2012, p. 1111). In order to obtain the highest possible number of articles associated with DSR, we used three metadata (topic, type of document, and language) as criteria for the WoS query. The research parameters used are described in the upper part of Fig. 6.1.

The query was executed in August 2018 and resulted in the identification of 156 articles. Skimming was used in the reading of these articles (Duggan & Payne, 2009) to identify applied DSR. For this purpose, two analysis codes were used: "name of the artifact" and "function of the artifact." 92 articles (59%) were identified as applied DSR. The excluded articles were classified into three categories: (a) 48 theoretical articles focusing on diffusing the DSRM; (b) 13 articles involving variations of the DSRM, 11 of which were associated with the Action Design Research (ADR) approach and 2 with the Secondary Design approach; and (c) 3 articles that merely cited Design Science, not as an approach but rather as a concept. The 92 articles identified as applied DSRM will from here on be referred to as the "original sample."

Identification of the Articles That Cited the 92 Articles Associated with Applied DSR

For the longitudinal analyses of the evolution of DSR within the concept of an evolutionary research project made up of various articles (the first premise described in the previous section), we sought to identify the articles that cited each of the 92 articles associated with applied DSR. To identify the citing articles, we used the Google Scholar search tool. For each of the 92 articles from the original sample, we made a search using a combination of two criteria: presence of the full name of the first author of the article and presence of the title of the article. The return screen for this search on Google Scholar presented the following information: title of the article, names of the authors, abstract, and a link entitled "Cited by" that is followed by a number that indicates the quantity of articles that

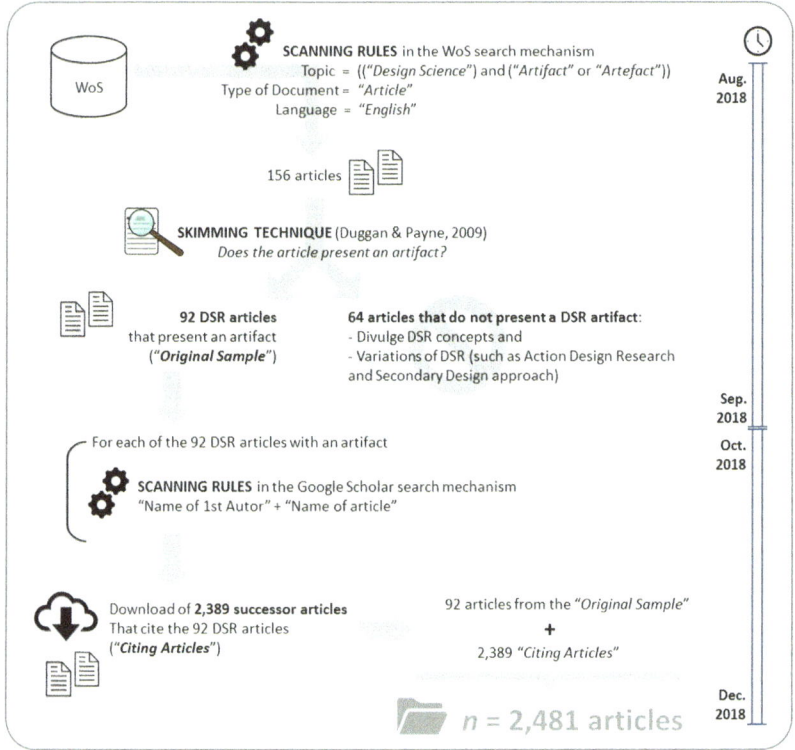

Fig. 6.1 Activities for identifying the analyzed articles

cited the article in question. By clicking on the "Cited by" link, a new tab opens containing the title, abstract and link to the site of origin of every article that cited the article from the original sample. By clicking on the link to the site of origin of each citing article, it is possible to download the text of the citing article or, when the article is not available due to restricted access, the metadata of the citing article can be collected to be used as search criteria in repositories of scientific articles. By following this procedure for the 92 articles from the original sample, we obtained a total of 2,389 citing articles. The search criteria, duration of time and quantity of results achieved through the activities for identifying the articles associated with DSR and for identifying the articles that cite them are described in Fig. 6.1.

Adopted Analysis Techniques

The identification of codes and themes for analyzing the texts of the articles followed the precepts of the content analysis technique (Miles & Huberman, 1994). The codes used to analyze the content of the 92 DSR articles were different from the codes used to analyze the 2,389 articles that cited the first 92 articles. We will begin by describing the procedures used for the content analysis of the 92 articles from the original sample.

Procedures for the content analysis of the 92 articles that presented DSR artifacts. The codes defined for the analysis were divided into two groups: the first with the descriptive data of the artifact; the second involving the data associated with testing the artifact. These two groups of codes are described in Table 6.1. The need to include the codes associated with testing the artifact in the 92 articles of the original sample was identified on the first reading of the articles. We observed that several of these articles only described and presented the artifact, with no mention of testing, in other words, they did not instantiate the artifact. We understand that, to the reader, instantiation can mean a greater evolution of the artifact and also that it will be easier to understand and, consequently, more likely to be reused by other researchers. Therefore, we included the codes "adequate user tester," "adequate environment for use" and "adequate inputs for use" for the purpose of comparing the pertinence of the tests conducted with the artifact proposed by the authors. When an artifact complied with these conditions, we classified it as having undergone an evaluation process. Artifacts that were used but did not meet the three conditions were classified as having undergone a demonstration process.

To analyze the texts of the articles, the researcher triangulation technique (Patton, 1987) was used. For this purpose, every article was read and coded at different times by two researchers (my colleagues) from a group of four experienced researchers with good knowledge of DSRM. The results of these analyses were then consolidated by me. The identification of diverging analyses was resolved, if necessary, with me and the two colleagues analyzing and discussing the conflicting points.

Procedures for the content analysis of the 2,389 citing articles. For each of the 92 articles from the original sample, a project was created using NVivo

Table 6.1   Codes for the content analysis of the articles from the original sample

	Code	Guiding question for analyzing the texts of the articles
Description of the artifact	Name of the artifact	What name is given to the artifact?
	Type of artifact	What type of artifact is it? Can one or more[a] of the following values be attributed to it: construct, model, method or instantiation (Hevner et al., 2004)
	Function of the artifact	What is the main purpose attributed to the artifact by its developers?
Test of the artifact	Use of the artifact	Was the artifact effectively used in its core function?
	Adequate user tester[b]	Was the artifact operated by a typical user?
	Adequate environment for use[b]	Was the artifact used in an appropriate environment of typical users?
	Adequate inputs for use[b]	Were data or other resources of the appropriate environment used for the tests?

[a]It may have multiple vales, considering that an article can describe a wide-ranging DSR study with the development of more than one artifact associated with the same function, for instance, a proposed method and its instantiation in one or more organizations
[b]For these three codes, the person responsible for using the artifact during the tests was observed, along with the location in which the tests were conducted and the inputs used during the tests

software for qualitative analysis. For each of these 92 projects registered in NVivo, the files referring to the articles from journals and conferences that cite them were uploaded, those identified by Google Scholar. With the purpose of directing and streamlining the process of reading and understanding the texts of the 2,389 citing articles, we initially used skimming to read the excerpts of texts that included the name of the artifact or the surname of the first author of the article from the original sample. It was observed that the vast majority of the citations (99.998%) were for diverse purposes and not related to the use of the artifact, such as: (a) mentioning the results or concepts addressed in the DSR article; (b) highlighting the existence of the artifact; (c) describing operational procedures or forms associated with the artifact; (d) describing activities involved in the development of the artifact; and (e) justifying the importance of the DSRM.

Table 6.2   Codes for the content analysis of the citing articles

Code	Guiding question for analyzing the texts of the articles
Use of the artifact	What evidence indicates the use of the artifact from the original sample?
Generation of theory	From the data associated with the use of the artifact, is there any proposition of theory or revision of theory?
Importance of the DSR artifact	How important is the DSR artifact in the process of generating theory?

From this initial reading of the citing articles, excerpts of texts that characterized actions involved in the use of the artifact from the original sample were identified, which were associated with the "use of the artifact" code described in Table 6.2. This was followed by the detailed and intensive reading of texts only in the articles that presented the use of the artifact from the original sample. The purpose of this reading was to identify the citing articles that proposed a generation of theory and the importance of using the artifact from the original sample to achieve this goal. To conduct this analysis, we used the codes "generation of theory" and "importance of the DSR artifact," described in Table 6.2. To analyze this second set of articles, the 2,389 cited articles, the researcher triangulation technique was also employed to resolve conflicting semantic interpretations.

FINDINGS

The content analysis of the texts of the 2,389 cited articles identified four that described the use of one of the 92 artifacts described in the articles of the original sample. In other words, 4.3% of the artifacts were (re)used by academics. Table 6.3 contains the results of the content analysis of the four articles from the original sample that were cited and whose proposed artifacts were (re)used in subsequent articles. Table 6.4 shows the results of the content analysis of the texts associated with the four subsequent articles, in other words, the articles that cite and use the four artifacts from the original sample. The content analysis of the texts of the four articles from the original sample indicated that in their proposition there was no formal evaluation process of the artifact. This information is shown

Table 6.3 Content analysis of the articles from the original sample whose artifacts were reused in later articles

Article that presents the DSR artifact [#Id article] + (authors, year)	Description of the artifact			History of the authors in the development of the artifact	Test of the artifact					Type of Test[b] Demonstration, evaluation or no test?
	Name	Type[a]	Function		Use of the artifact	Adequate user tester	Adequate environment for use	Adequate inputs for use		
[41] (Venkatesh et al., 2017)	"auto-ID enabled shopping assistance" ([41], p. 85)	[in]	"help in store customer (i.e., shopper) purchase decisions with auto-ID scanning functions that provide product descriptions and customer product reviews" ([41], p. 83)	In 2003, one of the authors studied user acceptance of information technology (Venkatesh et al., 2003), and in 2012, acceptance in a more specific context: that of the consumer (Venkatesh et al., 2012). These studies helped to provide the foundation for the constructs [co], in other words, the scales for evaluation used in the artifact described in [41]	Yes	Yes	No "Using a retail store laboratory, we conduct two empirical studies,..." ([41], p. 85)	No "We used product content that we downloaded from a major online retailer and we developed an application that displayed the content on shoppers' smartphones, such as Apple's iPhone" ([41], p. 89)		[dm]

Article that presents the DSR artifact [#Id article] + (authors, year)	Description of the artifact			History of the authors in the development of the artifact	Test of the artifact					Type of Test[b] Demonstration, evaluation or no test?
	Name	Type[a]	Function		Use of the artifact	Adequate user tester	Adequate environment for use	Adequate inputs for use		
[103] (Astor et al., 2013)	"game-based NeuroIS tool" ([103], p. 248)	[me] + [im]	"NeuroIS tool for improving emotion regulation" ([103], p. 247)	The authors published two years earlier (Adam et al., 2011) a framework for emotional bidding, describing the construct [co] used in the method [me] of the article described in [103]	Yes	No "that involved demonstration to real practitioners at trader and investor shows as well as play-testing calibration with student testers" ([103], p. 259)	No "The design artifact was then evaluated in two laboratory experiments" ([103], p. 248)	No "The design artifact was then evaluated in two laboratory experiments" ([103], p. 248)		[dm]

(continued)

Table 6.3 (continued)

Article that presents the DSR artifact [#Id article] + (authors, year)	Description of the artifact		Function	History of the authors in the development of the artifact	Test of the artifact				Type of Test[b]
	Name	Type[a]			Use of the artifact	Adequate user tester	Adequate environment for use	Adequate inputs for use	Demonstration, evaluation or no test?
[133] (Osterle & Otto, 2010)	"method for Consortium Research" ([133], p. 284)	[me]	"The method presented in the paper intends to propose a set of practices for researchers and practitioners collaborating in the design of IS artifacts." ([133], p. 284)	In the texts of the article itself, the authors indicate that they developed the construct [co] in previous articles: "The principles and results of Consortium Research have been published in scientific outlets (Osterle & Otto, 2009; Otto & Osterle 2010a, 2010b)" ([133], p. 286)	No "Based on a **self-evaluating** design process over a period of twenty years, the method's contribution is twofold." ([133], p. 292)	No [not used]	No [not used]	No [not used]	[nt]

| Article that presents the DSR artifact [#Id article] + (authors, year) | Description of the artifact | | | History of the authors in the development of the artifact | Test of the artifact | | | | | Type of Test[b] |
	Name	Type[a]	Function		Use of the artifact	Adequate user tester	Adequate environment for use	Adequate inputs for use		Demonstration, evaluation or no test?
[134] (Petter et al., 2010)	"framework for evaluating patterns" ([134], p. 9)	[mo]	"we propose a framework to evaluate patterns in any domain" ([134], p. 9)	Two years earlier, in Khazanchi et al. (2008), one of the authors had already developed the construct [co] to be used in the model, in other words, in the "framework for evaluating patterns"	No "we propose a framework to evaluate patterns in any domain and provide **examples of how to use** the evaluation framework" ([134], p. 9)	No [not used]	No [not used]	No [not used]		[nt]

[a]Legend [1]: [co] construct; [mo] model; [me] method; [in] instantiation
[b]Legend [2]: [dm] demonstration; [v] evaluation; [nt] no test

Table 6.4 Content analysis of the citing articles that use an artifact from the original sample

Article that cites the article of the DSR artifact [#Id article] + (Authors, Year)	Article that presents the cited DSR artifact [#Id article] + (Authors, Year)	Use of the artifact (in the article that cites the DSR artifact)	Generation of theory (in the article that cites the DSR artifact)	Importance of the DSR artifact (for the Generation of Theory)
[41-9] (Hoehle et al., 2019)	[41] (Venkatesh et al., 2017)	"Based on prior work on mobile application usability (Hoehle & Venkatesh, 2015; Hoehle, Zhang, & Venkatesh, 2015) and autoID-based research (Aloysius & Venkatesh, 2013; Venkatesh et al., 2017), we hypothesize that hardware design will have an effect on how users perceive autoID-based mobile application usability" ([41-9], p. 97)	**Occurred** A research model was prepared through testing and discussing hypotheses. "We found that adhering to mobile application usability principles could mitigate privacy concerns and consequently, improve shopping efficiency ([41-9], p. 91)	**Fundamental** The DSR artifact was the tool that enabled the action observed during the experiment, from which the data were collected and used in the tests

Article that cites the article of the DSR artifact [#Id article] + (Authors, Year)	Article that presents the cited DSR artifact [#Id article] + (Authors, Year)	Use of the artifact (in the article that cites the DSR artifact)	Generation of theory (in the article that cites the DSR artifact)	Importance of the DSR artifact (for the Generation of Theory)
[103-5] (Taubner et al., 2015)	[103] (Astor et al., 2013)	"we are particularly interested in bidders' immediate emotions (i.e., short-lived subjective experiences) (Rick & Loewenstein, 2008) in response to specific auction events (Astor et al., 2013) and the bidders' overall arousal (i.e., the intensity of the overall emotional state) during the auction process (Ku et al., 2005)" ([103-5], p. 841)	**Occurred** There is research model prepared from the test and discussion of the hypotheses. "… agency affected bidding behavior and its relation to overall arousal: whereas overall arousal and bids were negatively correlated when competing against human bidders, we did not observe this relationship for computerized agents. In other words, lower levels of agency yield less emotional behavior ([103-5], p. 838)	

(continued)

Table 6.4 (continued)

Article that cites the article of the DSR artifact [#Id article] + (Authors, Year)	Article that presents the cited DSR artifact [#Id article] + (Authors, Year)	Use of the artifact (in the article that cites the DSR artifact)	Generation of theory (in the article that cites the DSR artifact)	Importance of the DSR artifact (for the Generation of Theory)
[133-57] (Holler et al., 2017)	[133] (Österle & Otto, 2010)	"Through a consortium research approach (Österle & Otto, 2010), we had the opportunity to study case organizations in an intensive way from an inside perspective. In addition, we had the chance to incorporate cases from automotive innovation hubs like Singapore and Silicon Valley (Ebel and Hofer, 2016)" ([133-57], p. 119)	**Occurred** "the paper at hand (1) proposes, (2) describes, and (3) validates archetypes of e-collaboration for product development in the automotive industry" ([133-57], p. 114) "Considering the framework of Gregor (2006), classifications are equaled with 'type one theories' affording analysis and description without prediction. Hence, we can specify our theoretical contribution as 'theory for analysis'" ([133-57], p. 126)	**Secondary** The DSR artifact served as a tool to support communication and collaboration between the automotive industry specialists involved in collections through focus groups and semi-structured interviews

Article that cites the article of the DSR artifact [#Id article] + (Authors, Year)	Article that presents the cited DSR artifact [#Id article] + (Authors, Year)	Use of the artifact (in the article that cites the DSR artifact)	Generation of theory (in the article that cites the DSR artifact)	Importance of the DSR artifact (for the Generation of Theory)
[134-5] (Hoffmann et al., 2013)	[134] (Petter et al., 2010)	"To achieve our goal, we followed the design science based framework for patterns by Petter et al. (2010)" ([134-5], p. 2)	**Non-existent** The article [134-5] proposes a pattern for analyzing application websites: "To help providers select supporting measures for the website to improve the perceived trustworthiness of their application, we propose a set of requirement patterns" ([134-5], p. 1). As the artifact proposed by Petter et al. (2010) is a generalist model [mo] with criteria for the analysis and evaluation of patterns, its use has been highly appropriate. Thus, we have more generic artifact for analyzing criteria of patterns, used for the development of another more specific pattern (artifact)	**Non-existent** There was no generation of theory, but the development of another artifact (artifact generating artifact)

in the last column of Table 6.3, "Type of test," inferred from the consolidation of four codes associated with the super-code "Test of the artifact," which comprises the codes "use of the artifact," "adequate user tester" "adequate environment for use," and "adequate inputs for use." These analyses indicated that in two cases there was a demonstration process (Astor et al., 2013; Venkatesh et al., 2017) and in the other two articles (Österle & Otto, 2010; Petter et al., 2010) the artifact was not tested in any way. From here on, these four articles will be referred to by their identification codes, as shown in the first column of Table 6.3, respectively: articles [41], [103], [133] and [134]. At the end of this section, we will return to this question regarding the types of tests performed and their possible impacts on the (re)use and citation of the artifacts in subsequent articles.

Although all the subsequent (citing) articles effectively used an artifact from the original sample, as observed in the texts of the "use of the artifact (in the article that cites the DSR artifact)" code in Table 6.4, not all of them used it directly as an instrument to generate theory. This was shown in the analysis associated with the code "importance of the DSR artifact (for the generation of theory)," described in the last column of Table 6.4. In two of the four citing articles, the use of the artifact from the original sample can be observed playing a fundamental role in the generation of theory. This occurs in the article of Hoehle et al. (2019) and that of Teubner et al. (2015), from here on referred to, respectively, as Article [41-9] and [103-5]. The researchers and authors of these two citing articles monitored and recorded throughout their studies the data of the results obtained by using the DSR artifact. This enabled them to have all the necessary data for testing hypotheses associated with their theories. It should be highlighted that these two citing articles, which generated theory through the use of the DSR artifact, cited artifacts that had undergone the demonstration process, respectively, the artifacts described in Articles [41] and [103]. In these two articles that generated theories, using the artifact as a primary source for generating data, we observed that they had something in common: there is at least one author in common in both episodes, in the article that presents the artifact and the article that uses the artifact for the generation of theory. This aspect will be revisited and analyzed later in subsubsection "Evolutionary history of the four DSR projects with subsequent articles."

The other two articles, which presented the use of artifacts from the original sample, did not use them as a primary source for the generation of data for their research. The article by Holler et al. (2017), from here on identified as [133-57], which cited the artifact described in Article [133], did indeed present the development of a theory, but without using the data generated from the use of the DSR artifact. As highlighted in the "use of the artifact (in the article that cites the DSR artifact)" code in Table 6.4, the authors of Article [133-57] used the artifact described in Article [133], called the "consortium research approach," as a supporting tool for communication and collaboration between specialists in the automotive industry who participated in the focus group sessions and semi-structured interviews. The artifact was used as a tool to conduct the work of the researchers. Regarding the article of Hoffmann et al. (2013), from here on identified as [134-5], the authors used the artifact described in Article [134], not to generate theory but as a tool to create a second DSR artifact (artifact generating artifact). For this reason, it was given the denomination "non-existent" as a result of the analyses for the "importance of the DSR artifact (for the generation of theory)" code described in the last column of Table 6.4.

At this point of the analysis, in which we discuss the identification of studies that generated theory based on data obtained through the use of artifacts, we understand that it is important to highlight some situations other than the ones that are the object of this study and which have proved to be very difficult to differentiate in the first instance. A very common situation was that of the coupling (and non-use) of artificial objects to generate another artificial object, in other words, an artifact helping to build another artifact. Gregor and Hevner (2013, p. 341) refer to these situations as a DSR contribution at "Level 2 in the form of nascent design theory (e.g., constructs, design principles, models, methods, technological rules)." An example was that of the article from the original sample by Piirainen et al. (2012). In this article, a construct type artifact is presented ("Synthesis of the challenges for collaborative design"), which was later used by Bittner and Leimeister (2014) for the development and instantiation of a method type artifact ("MindMerger"). In this case, there are citations of the artifact from the original sample, but not with regard to the specific use of the artifact, but rather incorporating some of its concepts and abstractions for the generation of a new artifact. It is important to note that the cycle of nascent design theory very often involves the intertwining and composition of different types of artifacts (construct, model, method and instantiation) over time, in different episodes and very often in diverse manuscripts. Therein lies the importance of accurate research and with the distinction of this situation in relation to the one that interests us: the generation of theory from data gathered through the use of a DSR artifact. For this purpose, it is fundamental to have a correct understanding and identification of the codes "type (of artifact)" and "function (of artifact)," both of which are presented and described in Table 6.3.

Scope of the Test Conducted Using the DSR Artifact and Its Impact on Its Citations and (Re)use in Subsequent Articles

Analyzing the type of test conducted on the 92 artifacts from the original sample, we identified three categories of tests: (a) 13 articles (14%) effectively evaluated the artifact, in other words, it was instantiated and used by typical users in its natural use environment and with data from the real use environment; (b) 67 articles (72%) demonstrated the artifact, in other words, they instantiated and used the artifact, but without meeting all the requirements for evaluation (use by a typical user, testing in a typical environment and use of typical inputs); and (c) 12 articles (13%) only described and presented the artifact without conducting any type of test. These results proved to be in keeping with the perceptions of Baskerville et al. (2015, p. 560): "much of the current empirical work in design-science actually engages science in an idiographic way, often involving some form of prototyping." It is important to observe that the term prototyping denotes a stage of testing the artifact that is highly incipient. Using the terminology and the phases of DSR according to Peffers et al. (2007), the set of 92 DSR articles that were analyzed are far more consistent and pertinent concerning the actions and results of the demonstration phase than the evaluation phase of the DSR artifact.

Regarding the 92 articles from the original sample, there was a strong and positive association between the time of publication and the volume of subsequent citations (regression analysis: number of citations received versus the year of publication of the article, p-value: 0.000). Another relational analysis that was conducted was between the citations received by the articles from the original sample and the types of tests (demonstration, evaluation or no test) performed on the artifacts. For this latter analysis, the non-parametric Kruskal-Wallis statistical test for K independent samples was used due to the normality indicated by the Komogorov-Smirnov test. The Kruskal-Wallis test indicated that there is no effect between the groups (demonstration, evaluation or no test) on the citations ($\chi^2(2) = 4.049$; $p = 0.132$). This indicates that the three groups had equivalent behavior in relation to the means. Thus, it was observed that all the DSR articles from the original sample were equally read and perceived by academics in terms of citations, irrespective of the type of test performed with the artifact.

The citations investigated during the data collection show that 92 articles from the original sample were identified and recognized by the scientific community. This makes the identified context more intriguing, considering that the premises of DSR define its artifacts as something useful, innovative and intended for a well-defined type of problem and public. Therefore, it is to be expected that among the diverse researchers who read and cited these articles, some could conceive opportunities for the continuity of these studies associated with the use of these innovative artifacts. From the perspective of the practitioner (the one who uses the artifact) and the beneficiary of the artifact (the one affected by the actions of the artifact) a number of opportunities for analysis may be considered. The perspective of grounded theory is one of these alternatives, with researchers able to explore the impact of the inclusion of the new (artifact) in a certain environment, in other words, "how the actors respond to changing conditions and to the consequences of their actions" (Corbin & Strauss, 1990, p. 5).

An analysis of the (re)use of artifacts from the original sample from the perspective of those who made it is jeopardized because of the low number of occurrences: only 4 situations were identified. However, we can explore another perspective of this same phenomenon: the non-(re)use of most of the artifacts (88) from the original sample. A possible cause of the few cases of (re)use of the artifacts by researchers might be that the tests lacked development and maturity. As observed, the tests were insufficient for 85% of the articles from the original sample: 13% of the articles did not include tests and 72% merely demonstrated the artifacts. It is important to remember that for academics the use of superficial tests on artifacts is associated with scientific knowledge that has not been very well developed (Baskerville et al., 2015). Therefore, the deficiencies observed in the process of testing the artifacts from the original sample may have hindered the understanding of the artifact and had a negative influence on the perception of academics regarding the quality and utility of the artifacts. These factors may have been part of the reason for the few situations of (re)use of the artifacts among academics.

Evolutionary History of the Four DSR Projects with Subsequent Articles

An analysis of the four citing articles that made use of the articles described in the original sample ([41], [103], [133], [134]), showed that in two cases ([41] and [103]) the articles from the original sample and the citing articles have authors in common. In these two subsequent articles with authors in common for the article from the original sample and the citing article, the authors used the data obtained from the reuse of the DSR artifact as a primary source for testing the hypotheses and for the generation of theory. The researchers Viswanath Venkatesh, John Aloysius, and Hartmut Hoehle are the authors of both Article [41], which presented a DSR artifact, and the subsequent article [41-9], which cites the use of the artifact. Likewise, Marc Adam is one of the authors of Article [103] from the original sample and the subsequent article [103-5], which cites and uses the same artifact.

The identification of common authors for the articles from the original sample and the subsequent (citing) articles led us to investigate the history of authorship in a larger context: publications prior to the article from the original sample. To this end, we researched the previous works of the authors of the four articles from the original sample in which we identified the (re)use of their artifacts. All of the authors of these four articles had also made an initial contribution in the form of a proposition for an artifact of the construct type in previous articles. This abstract artifact was coupled and used in the development of the following DSR artifact, which we identified in the original sample. The history of this prior contribution can be observed in the code "history of the authors in the development of the artifact" in Table 6.3. The following two paragraphs describe the longitudinal time trajectory of articles and types of artifacts for these two DSR projects. We will address the two articles directly associated with the focus of this study, in other words, those whose use of their artifacts enabled the generation of data that were used by the researchers to conduct tests and develop theories.

One of the developers of the artifact presented in Article [41] of the original sample, "auto-ID enabled shopping assistance," identified as the instantiation type, was the researcher Viswanath Venkatesh. This artifact was used as a source for generating data for the generation of theory in an article published two years later, in the subsequent article [41-9], of which the researcher Viswanath Venkatesh is the fourth author. It is important to note that the involvement of Viswanath Venkatesh with the themes and core concepts of the DSR artifact in Article [41] goes back a long way. Fourteen years previously, in the first episode of this DSR project, Venkatesh developed and published an artifact of the construct type (Venkatesh et al., 2003), which aided the foundation of the artifact developed and presented in the second episode (Article [41]). This article would generate the data necessary for the tests and the generation of theory found in the third episode [41-9]. Venkatesh's actions leading up to the artifact presented in Article [41] are described in the code "history of the authors in the development of the artifact," in Table 6.3. It was observed that Venkatesh was involved in this DSR project for a period of 16 years. On three occasions or distinct episodes, three articles were produced, encompassing the development of two types of DSR artifacts (construct and instantiation) and a theory.

The other DSR artifact, whose data stemming from its use were used for tests and the generation of theory, is the "game-based NeuroIS tool." This artifact was developed and presented in Article [103], with the researcher Marc Adam as one of its authors. The data obtained through the use of this artifact were used to conduct tests and generate theory in an article published two years later, Article [103-5]. In this second episode of the DSR project, characterized by Article [103-5], the researcher Marc Adam is the second author. Thus, as in the relationship of Venkatesh with the DSR project and the artifact associated with Article [41], the relationship between Marc Adam and the DSR project and artifact associated with Article [103] stems from before the publication of this article from the original sample. In the "history of the authors in the development of the artifact" code, in Table 6.3, it can be seen that two years earlier the researcher Marc Adam had already published studies addressing the construct type artifact (Adam et al., 2011) that paved the way for the artifact presented in Article [103]. Therefore, we have a six-year trajectory of Marc Adam working on the DSR project, with three articles and the development of three types of artifacts (construct, method and instantiation) and a theory.

The data obtained from analyzing different levels and moments of the articles that present DSR artifacts, involving both their preceding and succeeding articles evidence and corroborate some of the recent statements of some of the authors that are renowned with regard to the DSRM. For example, the two DSR projects that also contributed to the development of a theoretical field did so in three different episodes (articles published over the years) and involving a sequencing of different types of artifacts (construct, method, model and instantiation). This corroborates the characterization of the DSRM according to Baskerville et al. (2015), who claimed that the knowledge generated by DSR is polymorphic and produced in different moments or episodes. These findings are also in keeping with the three levels of contribution of DSR described by Gregor and Hevner (2013). A discussion of the data linked to the contribution of the DSRM to the theoretical field, the object of this study, is given in the following section. This also implies the complementation of propositions recently formulated by renowned authors in the field of DSR.

DISCUSSION

Although the artifacts in the 92 articles from the original sample attracted the attention of academics, as characterized by the 2,389 identified citations, the (re)use of these artifacts was minimal, a phenomenon only observed for four of the 92 articles from the original sample: [41], [103], [133] and [134]. A greater restriction was observed for the (re)use of artifacts with the purpose of developing the theoretical field, a scenario in which the data stemming from the operation of the artifact are used for tests and theoretical discussions (Level 3 of the contributions of DSR). Only two occurrences of this phenomenon were identified associated with the (re)use of the artifacts described in Articles [41] and [103] of the original sample. The longitudinal analysis of the development of these two DSR projects, which culminated in the generation of theory, evidenced the heavy involvement of the researchers in the development of the artifact and the evolution of the DSR project. This involvement is significant, both with regard to time, in terms of years of dedication, and effort toward integration and synergy between the different types of artifacts (polymorphism) developed and presented in different episodes and articles.

It is important to highlight that the two articles from the original sample that were (re)used in subsequent articles for the generation of theory, articles [41] and [103], were not submitted to consistent tests in accordance with the precepts of the DSRM, as their artifacts were only demonstrated. The lack of a formal evaluation process for an artifact, which would better characterize the space of the problem and the proposed solution, implies less information and fewer discussions regarding the artifact and, consequently, greater difficulty for other researchers to reuse the artifact and proceed with the DSR project. In these two DSR projects, which reached contribution Level 3, this difficulty was overcome by the engagement and continuous presence of a researcher throughout the project's cycle. For these DSR projects, one author participated in the three articles associated with the three episodes of research: (i) article from the original sample, which presented the artifact; (ii) preceding article that developed a construct used for the development of the artifact; and (iii) subsequent article that cites the use of the artifact to generate data, tests, and a discussion of theory. In the research project associated with the artifact described in Article [41], the researcher Venkatesh is the author present in all three moments. Likewise, in Article [103], Adam is the researcher who is present at all three moments. These researchers played the role of integrators, accumulating and sharing with the different teams the knowledge obtained over the diverse episodes of the DSR project. Thus, they may have helped the teams of researchers that worked on the different episodes of the DSR project to overcome the difficulties involving a lack of information resulting, for example, from superficial test processes of the artifacts.

In addition to the sharing of authors in different episodes of DSR projects, we should highlight the differences, in other words, the different authors involved in the different episodes. In Article [103], 5 authors are involved, while in the subsequent article, [103-5], there are three authors. The only author in common between these two articles is the researcher Marc Adam, who is the second author of both articles. The first author and other research partners are totally different from one article to the other. A comparative analysis of the authorship of Article [41] from the original sample and the subsequent article that cites and uses its artifact, Article [41-9], identified a different author in each of the articles: Scot Burton is an author only of [41], whereas Soheil Goodarzi is an author only of Article [41-9]. Both articles have 4 authors, in other words, three authors in common, but with a different order of authorship. In Article [41], Venkatesh was the first author, while in Article [41-9] he was the fourth. In Article [41-9], Hoehle was the first author, while in Article [41] he was the third.

The alterations made to the DSR project teams, both in terms of intensity of involvement, characterized by the order of appearance on the list of authors, and different members should be considered normal, and even expected in extensive DSR projects that reach Level 3 of contribution. This is justified considering the different competencies required throughout the DSR project. The competencies required for the development of the scientific artifact associated with a Level 1 contribution of DSR (idiographic science), within a pragmatic research paradigm (Creswell, 2014), are not the same as those required, for example, for the development of theory at Level 3 (nomothetic science). This perception is relevant to the discussion of the characterization of DSR projects. The characterizations of DSR projects according to Baskerville et al. (2015) were confirmed: polymorphic knowledge production, in other words, encompassing different types of artifacts, generated in different moments or episodes and with the writing and publication of diverse manuscripts (articles). From the findings of this study we can make an addition to the set of statements regarding the characterization of DSR projects: the required participation of different actors (researchers), with different competencies and levels of involvement throughout the trajectory of the DSR project.

The two DSR projects that made theoretical contributions, associated with Articles [41-9] and [103-5], based their tests on data obtained from using the artifacts that were heavily dependent on resources of communication and information technology. These two theoretical contributions were published in journals in the field of information systems. However, their discussions could lead to publication in journals in other fields that are equally or more pertinent with regard to the themes addressed by theories. Article [41-9] addresses themes of interest to the field of Marketing, while Article [103-5] addresses themes of interest to the field of Psychology. Forwarding for publication in the field of information systems must be due to the profile of the researchers who acted as integrators of the two DSR projects, both professors in the field of information systems. Thus, the current context of continuity of the DSR projects is limited regarding the target audience, restricted to the same profile as the authors. Of the 92 articles from the original sample, 78 (85%) were published in journals in the field of Information systems. Thus, a challenge for the current context of DSR lies in making the technological artifacts available to other audiences outside the field of their developers, with other theoretical perspectives and possible interest in extrapolating the DSR contributions from Level 1 (idiographic) to Level 3 (nomothetic).

The reflections contained in this chapter help to overcome this limitation. The differentiation between the types of tests of DSR artifacts and the opportunity for theorization based on the data stemming from the use of the artifact help to diffuse a broader view of DSR projects. Thus, reviewers and authors can view the performance of evaluation type tests as an opportunity to give continuity to articles that only demonstrated their artifacts. The justifications for this include proving the utility of the artifact within a realistic context and the possibility of theorizing by analyzing how the actors (practitioners and beneficiaries of the artifact) respond to changing conditions and to the consequences of the introduction of the new artifact. In terms of actions of an administrative order, editors can suggest the inclusion of two specific metadata to identify and classify articles that present DSR artifacts: (a) the community of practitioners for whom the artifact was projected (target audience); and (b) potential clients or beneficiaries of the actions of the artifact. Highlighting this information as metadata, in other words, elements that identify and classify, increases the chance of visibility for DSR articles in other research communities.

The trend to conduct DSR is positive with regard to improving the testing of artifacts, considering the natural maturity process of recently created research strategies. As examples of this trend, the studies that focus on the evaluation process of DSR artifacts recently conducted by Baskerville et al. (2015) and Venable et al. (2016) may be highlighted. This new context should make the testing of artifacts more rigorous and complete. The improved testing of artifacts, going beyond the demonstration phase, should improve researchers' perception and understanding of the artifact. This can encourage researchers to consider actions such as applying innovative artifacts in a determined equally innovative context, configuring a scenario hitherto inaccessible for scientific observation. In this case, we would have a proposal for theories based on revelatory cases (Yin, 2014). Movements like this can raise the contributions of DSR studies to Level 3, in other words, helping to generate theories.

Thinking of DSR also as a source of generating theory, it is important to revisit the initial texts that disseminated the DSRM. Romme (2003) indicated that the development of artifacts within the principles of the design mode could be inspired by the empirical findings of the science mode. The findings of this chapter show that the reverse is also possible, that empirical findings of the science mode can also be inspired by and grounded in data and events associated with the use of artifacts produced by the design mode. Thus, there is a recursive collaboration between science mode and design mode. With this understanding of theorization as one of the possible contributions of DSR, some explanatory models of DSR could be broadened and complemented, such as the one associated with the possibility of entry points for the continuity of DSR projects proposed by Peffers et al. (2007). The new entry point to be added originates from the evaluation phase of the artifact and presents the theorization phase as its destination, based on the behaviors of those involved, observed through use of the artifact.

In this chapter, it was established that the scope of discussions on the impact of DSR artifacts would be restricted to the academic/scientific community, considering that the objective was to analyze the potential of artifacts to generate theories (Level 3 contribution). Therefore, here we only addressed the scientific repercussions of DSR artifacts without considering all the other typical audiences of a study within the pragmatic research paradigm (Creswell, 2014). A broader analysis of the impacts of artifacts in terms of society could also include an analysis of comments on social media and in specialized media used by practitioners and managers in fields directly associated with the artifact, as proposed by Hevner et al (2004).

References

Adam, M. T. P., Krämer, J., Jähnig, C., Seifert, S., & Weinhardt, C. (2011). Understanding auction fever: A framework for emotional bidding. *Electronic Markets, 21*(3), 197–207.

Aloysius, J. A., & Venkatesh, V. (2013). *Mobile point-of-sale and loss prevention: An assessment of risk*. Retail Industry Leaders Association [WWW document]. http://waltoncollege.uark.edu/lab/JAloysius/RILA%20Report/MobilePOSReport.pdf.

Astor, P. J., Adam, M. T. P., Jercic, P., Schaaff, K., & Weinhardt, C. (2013). Integrating biosignals into information systems: A NeuroIS tool for improving emotion regulation. *Journal of Management Information Systems, 30*(3), 247–277.

Baskerville, R. L., Kaul, M., & Storey, V. C. (2015). Genres of inquiry in design-science research: Justification and evaluation of knowledge production. *MIS Quarterly, 39*(3), 541-A9.

Bittner, E. A. C., & Leimeister, J. M. (2014). Creating shared understanding in heterogeneous work groups: Why it matters and how to achieve it. *Journal of Management Information Systems, 31*(1), 111–144.

Corbin, J., & Strauss, A. (1990). Grounded theory research: Procedures, canons, and evaluative criteria. *Qualitative Sociology, 13*, 3–21.

Creswell, J. W. (2014). *Research design: Qualitative, quantitative, and mixed methods approaches* (4th ed.). Sage.

Duggan, G. B., & Payne, S. J. (2009). Text skimming: The process and effectiveness of foraging through text under time pressure. *Journal of Experimental Psychology Applied, 15*(3), 228–42.

Ebel, B., & Hofer, M. B. (2016). *Automotive management—Strategie und marketing in der automobilwirtschaft* (2nd Ed.). Berlin/Heidelberg, Germany: Springer.

Glaser, B. G., & Strauss, A. L. (1967). *The discovery of grounded theory: Strategies for qualitative research*. Aldine de Gruyter.

Gregor, S. (2006). The nature of theory in information systems. *MIS Quarterly, 30*(3), 611–642.

Gregor, S., & Hevner, A. (2013). Positioning and presenting design science research for maximum impact. *MIS Quarterly, 37*(2), 337–355.

Guba, E. G. (1981). Criteria for assessing the trustworthiness of naturalistic inquiries. *Educational Communication and Technology, 29*(2), 75–91.

Hevner, A. R., et al. (2004). Design science in information systems research. *MIS Quarterly, 28*, 75–105.

Hoehle, H., Aloysius, J. A., Goodarzi, S., & Venkatesh, V. (2019). A nomological network of customers' privacy perceptions: Linking artifact design to shopping efficiency. *European Journal of Information Systems, 28*(1), 91–113.

Hoehle, H., & Venkatesh, V. (2015). Mobile application usability: Conceptualization and instrument development. *MIS Quarterly, 39*(2), 435–472.

Hoehle, H., Zhang, X., & Venkatesh, V. (2015). An espoused cultural perspective to understand continued intention to use mobile applications: A four-country study of mobile social media application usability. *European Journal of Information Systems, 24*(3), 337–359.

Hoffmann, A., Hoffmann, H., & Söllner, M. (2013). Fostering initial trust in applications—Developing and evaluating requirement patterns for application websites. In *21st European Conference on Information Systems* (ECIS), Utrecht, The Netherlands.

Holler, M., Uebernickel, F., & Brenner, W. (2017, June 5–10). *Defining archetypes of e-collaboration for product development in the automotive industry*. Paper presented at the 25th European Conference on Information Systems (ECIS), Guimarães, Portugal. https://aisel.aisnet.org/ecis2017_rp/8

Khazanchi, D., Murphy, J. D., & Petter, S. C. (2008, May 23–24). *Guidelines for evaluating patterns in the IS domain*. Paper presented at the Third Midwest United States Association for Information Systems (MWAIS), Eau Claire, Wisconsin. https://digitalcommons.unomaha.edu/isqafacproc/7.

Kogut, B., & Zander, U. (1997). Knowledge of the firm: Combinative capabilities, and the replication of technology. In L. Prusak (ed.), *Knowledge in organizations* (pp. 17–35). Boston: Butterworth-Heinemann.

Ku, G., Malhotra, D., & Murnighan, J. K. (2005). Towards a competitive arousal model of decision-making: A study of auction fever in live and Internet auctions. *Organizational Behavior and Human Decision Processes, 96*(2), 89–103.

Kuechler, B., & Vaishnavi, V. (2008). On theory development in design science research: Anatomy of a research project. *European Journal of Information Systems, 17*(5), 489–504.

Merton, R. K. (1949). *Social theory and social structure: Toward the codification of theory and research*. Free Press.

Miles, M. B., & Huberman, A. M. (1994). *Qualitative data analysis: An expanded sourcebook* (2nd ed.). Sage Publications.

Österle, H., & Otto, B. (2009). *A method for consortial research. Institute of Information Management*, University of St. Gallen, Arbeitsbericht Nr. BE HSG/CC CDQ/6.

Österle, H., & Otto, B. (2010). A method for researcher-practitioner collaboration in design-oriented IS research. *Business & Information Systems Engineering, 2*(5), 283–293.

Otto, B., & Österle, H. (2010a). Practical relevance through consortium research? Findings from an expert interview study. In *5th International Conference on Design Science Research in Information Systems and Technology (DESRIST 2010)*, St. Gallen, Switzerland.

Otto, B., & Österle, H. (2010b). Relevance through consortium research? A case study. In *18th European Conference on Information Systems (ECIS 2010)*, Pretoria.

Patton, M. Q. (1987). *How to use qualitative methods in evaluation*. Sage.

Peffers, K., Tuunanen, T., Rothenberger, M. A., & Chatterjee, S. (2007). A design science research methodology for information systems research. *Journal of Management Information Systems, 24*(3), 45–77.

Peffers, K., Tuunanen, T., & Niehaves, B. (2018). Design science research genres: Introduction to the special issue on exemplars and criteria for applicable design science research. *European Journal of Information Systems, 27*(2), 129–139.

Petter, S., Khazanchi, D., & Murphy, J. D. (2010). A design science based evaluation framework for patterns. *Data Base for Advances in Information Systems, 41*(3), 9–26.

Piirainen, K. A., Kolfschoten, G. L., & Lukosch, S. (2012). The joint struggle of complex engineering: A study of the challenges of collaborative design. *International Journal of Information Technology & Decision Making, 11*(6), 1087–1125.

Rick, S., & Loewenstein, G. (2008). The role of emotion in economic behavior. In M. Lewis, J. M. Haviland-Jones, & L. F. Barrett (Eds.), *Handbook of emotions* (3rd ed., pp. 138–156). New York: The Guilford Press.

Robinson, O. C. (2011). The idiographic/nomothetic dichotomy: Tracing historical origins of contemporary confusions. *History & Philosophy of Psychology, 13*(2), 32–39.

Romme, A. G. L. (2003). Making a difference: Organization as design. *Organization Science, 14*(5), 558–573.

Teubner, T., Adam, M., & Riordan, R. (2015). The impact of computerized agents on immediate emotions, overall arousal and bidding behavior in electronic auctions. *Journal of the Association for Information Systems, 16*(10), 838–879.

Venable, J., Pries-heje, J., & Baskerville, R. (2016). FEDS: A framework for evaluation in design science research. *European Journal of Information Systems, 25*(1), 77–89.

Venkatesh, V., Aloysius, J. A., Hoehle, H., & Burton, S. (2017). Design and evaluation of auto-id enabled shopping assistance artifacts in customers' mobile phones: Two retail store laboratory experiments. *MIS Quarterly, 41*(1), 83–113.

Venkatesh, V., Morris, M. G., Davis, G. B., & Davis, F. D. (2003). User acceptance of information technology: Toward a unified view. *MIS Quarterly, 27*(3), 425–478.

Venkatesh, V., Thong, J. Y. L., & Xu, X. (2012). Consumer acceptance and use of information technology: Extending the unified theory of acceptance and use of technology. *MIS Quarterly, 36*(1), 157–178.

Yin, R. K. (2014). *Case study research: Design and methods* (5th ed.). Sage.

CHAPTER 7

Variations of the DSR Approach

In this chapter we will explore the incorporation of the precepts of the DSR approach alongside other research strategies, i.e., new methodological approaches resulting from the triangulation of DSR with other research strategies. We will first explore the integration between the DSR and the Action Research approach, which resulted in Action Design Research. Following this, we will explore the triangulation between the DSR and Grounded Theory, resulting in the Grounded Design approach.

ACTION DESIGN RESEARCH (ADR)

The ADR method is a research strategy derived from the triangulation between DSR and Action Research (AR) methods. Before describing the ADR method and its differential aspects in relation to DSR, we will make a brief presentation of the AR approach. Considering all exposition of DSR already done in the previous chapters, we will discuss the advantages of incorporating some of the characteristics of AR for the DSR strategy. In the opposite direction, that is, for those who only know AR, we will also develop a text of complementarity and gains for the AR approach when incorporating the assumptions of the DSR approach.

© The Author(s), under exclusive license to Springer Nature Switzerland AG 2021
J. O. De Sordi, *Design Science Research Methodology*,
https://doi.org/10.1007/978-3-030-82156-2_7

Action Research (AR)

German psychologist and social scientist Kurt Lewin coined the term AR in 1946, defining it as "a comparative research on the conditions and effects of various forms of social action and research leading to social action" (Lewin, 1946, p. 35). RA is social, empirical research of a participatory nature between researchers and members of the community associated with the situation or social problem it aims to address. The strong cooperation between researchers and members of the social community involved with the problem is one of the main characteristics of AR. The basic idea is to solve problems by assisting and promoting the social community directly associated with the problem. To this end, members of the community are involved from the very first moments of the research, when the problem and the objective of the research are defined. Thus, in terms of the philosophy of science, the AR strategy fits as a transformative research paradigm. According to Creswell (2014) the transformative research paradigm is characterized by being: political, power, and justice-oriented, collaborative, and change-oriented.

The central feature of the AR approach activities is the action-reflection cycle which is broken down into four phases: action planning, action execution, observation of results, and reflection on what can be changed in order to have a new version of the action that is more effective. This interactive process of the four phases of plan-act-observe-reflect, followed by other sets of four phases, is a system of learning by doing. Action is refined over time by approximation through practical experience of the actions undertaken by the research group of practitioners and community members. The presence of participants involved from the beginning of the research gives it more legitimacy, ensuring the broad and unrestricted interaction of community members. The continuous action-reflection cycle is named the Lewinian spiral after its mentor.

ADR Strategy

Like the DSR, the ADR strategy is characterized as a pragmatic research paradigm (Creswell, 2014, p. 39), that is, "concern with applications-what works-and solutions to problems." In DSR the delivery of relevant actionable knowledge to the scientific community and, in particular, to practitioners, takes place through a new artifact (Hevner et al., 2004; Romme, 2003). The main difference of the ADR method with respect to the DSR method is in the greater diversity and involvement of actors (researchers, practitioners, and users) during the development and evaluation of the artifact. The ADR method proposes several interactive cycles among the actors, according to the precepts of the Action Research (AR) method (Susman & Evered, 1978). The group of professionals for a research project with ADR method should be composed of researchers and practitioners, and the latter should bring information to the group from interactions with end-users in the field, when making use of the artifact. Thus, ADR is the result of a triangulation of techniques present in the DSR and AR methods.

According to Sein et al. (2011) the ADR method consists of four phases: Phase 1, Problem formulation; Phase 2, Building, intervention, and evaluation; Phase 3, Reflection and learning; and Phase 4, Formalization of learning. For each of these four phases, Sein et al. (2011) defined seven guiding principles for these four phases, described in the following paragraph.

For Phase 1, "Problem formulation," two principles were defined: (a) "practice-inspired research," which implies that researchers perceive the opportunity to generate knowledge that can be applied to the class of problems that the specific problem exemplifies; and (b) theory-ingrained artifact, which emphasizes that the ensemble artifacts created and evaluated via ADR are informed by theories. For Phase 2, "Building, intervention and evaluation," three principles were defined: (a) "reciprocal shaping," here there is a two-way street, community characteristics should be considered to shape the design of the new artifact in order for it to be simpler and easier to use by community members, as design characteristics should be chosen to elevate and shape actions and behavior of its future users (promote change); (b) "mutually influential roles," the shaping of the previous principle does not occur alone, researchers and members of the future user community (the practitioners) need to mutually exchange their knowledge, respectively, knowledge of advanced theories and technologies, and knowledge of group working practices; and (c) "authentic and concurrent evaluation," each new action change is immediately implemented and tested, according to the action-reflection cycle assumption. For Phase 3, "Reflection and learning," Sein and his colleagues defined the principle of "guided emergence." The idea of this principle is to highlight that the routine concept of design as something totally conceived by the designer in the initial phase and with few subsequent changes does not apply in ADR. The design is conceived in practice, in the interaction between researchers and practitioners, i.e., at each round of the action-reflection cycle, i.e., driven not by the original premises of the design, but by the observations and use of the artifact in the field. Finally, in Phase 4, "Formalization of learning," the principle of "generalized outcomes" which involves generalizing: (i) the problem, in terms of clearly identifying a class of problem; (ii) the solution, in terms of defining an artifact capable of solving the problem; and (iii) the principles of the design process, allowing the construction and improvement of the artifact by others.

An interesting aspect of the texts that describe the results of ADR is to try to give more understanding and clarity to the research outputs. The ADR for having a more social aspect (from the RA) is associated with researchers with less knowledge and domain of issues related to design, unlike the researchers of the DSR that has its origin in the engineering, more affectionate to design issues. Thus the explicitness of the ADR outputs seem to be a very important contribution even to those who practice the DSR. In the fourth and last phase of ADR, Phase 4, "Formalization of learning," there is the demand to present the meta specification of the artifact, making it easier for other researchers to understand and evolve the artifact, or even to conceive another artifact proposal for that class of problem. The meta specification is the central feature of Design Theory, which is composed by meta-requirements and meta-design. The meta-requirements are defined from a kernel of theories from natural or social sciences that govern the requirements of the artifact design (Walls et al., 1992). In this way, ADR brings to the practitioners of DSR not only the idea of inserting as soon as possible the practitioners (future users) in the development project of the new artifact, but also a more explicit and clearer form of the specifications of its outcomes.

Implications of ADR for RA Practitioners and Specialists

The seminal text presenting ADR, by Sein et al. (2011), as well as the vast majority of subsequent ADR articles, was conceived by researchers in the area of Information Systems from a technological perspective, grounded, and written from the perspective of practitioners of the DSR. Thus, we have that the audience of researchers with tradition in AR research are the most distant and least attended by the texts that present and describe the ADR approach. When considering that ADR is a recent approach with increasing adoption (Mullarkey & Hevner, 2019), it becomes notable the importance of also exploring the contributions and possible mishaps in understanding this approach from the perspective of practitioners and experts in the AR approach. This requires the development of analogies between AR and DSR.

The practitioners of the DSR approach recognize that the foundation of RA that most complements and enhances the DSR approach, when comparing the two approaches, is the co-creation that occurs between researcher and practitioner. This joint work brings greater understanding of the demand or the problem to be solved, including social and cultural issues, less perceptible to researchers external to the group of practitioners. Within this perspective of complementarity between approaches, it is also relevant to ask what would be the main benefits for RA researchers when incorporating fundamentals from the ADR approach. In the next paragraphs, some of these increments to the RA practitioners are addressed from the discussion of the DSR fundamentals inserted in the ADR approach.

In DSR the object and the central focus of the research is the artifact that should be demonstrably useful. On the other hand, AR research focuses on the intervention that may even result in an artifact, understood as something relevant, but secondary. In "AR the artefact is normally the by-product of the research intervention, not the goal of that intervention" (Papas et al., p. 156). The emphasis of RA is on solving local practical problems through the intervention, characterizing the result generated by the research as a local practice contribution, unlike the DSR that aims at general practice contribution through the artifact (Goldkuhl, 2013, p. 3). At the core of the contributions of the DSR for the mental model of the researcher with experience in the AR approach is the incorporation of the canons and assumptions of design theory, more specifically in thinking about the generalization of knowledge from the design of an artifact. The typology of artifacts recognized by the DSR approach, comprising method, construct, model, and instantiation (Hevner et al., 2004), proposes and favors the researcher to think and perceive different possibilities of contributions in the form of artifact.

The learnings from the artifact design process are formalized in meta-requirements and meta-designs. These meta-specifications facilitate the actions of critical analysis and evolution of the proposed artifact, as well as the conception and proposition of new artifacts for the same class of problem. In short, it goes far beyond ascertaining and allowing the replication of the solution. Many times, due to space limitation of the amount of words or even due to ontological challenges, DSR articles present only the artifact, without describing the meta-specifications. In Table 5.1 we present: (a) the meta-requirements, describing what a given artifact should do; (b) the meta-designs, describing the design features to meet the meta-requirements; and (c) the contextualization of the two previous aspects in the context of the design process traversed by the team of researchers and practitioners who developed and evaluated the artifact. Thus, in the DSR we also have the availability of contents associated with the artifact that allow to highlight and exemplify some of the main actions linked to the canons of design theory.

As to the artifact evaluation process, in RA one seeks to solve the locality problem, not necessarily with a solution notoriously recognized as the best among possible. In the DSR, on the other hand, there is the search for a solution that can be considered the best, either through a unique and innovative artifact or through an artifact similar to the current ones, but with superior performance. For needing to meet these requirements, the DSR has an artifact evaluation process more formal and structured than the one observed in the RA. In this sense, Papas et al. (2012) pointed out that in the DSR the evaluation process is vital, while in the RA this process is perceived as useful (Papas et al., 2012). While in the RA evaluation process the action learning is practiced, through an action-reflection cycle in a locality, in the DSR the learning from failures strategy is practiced, generating several successive versions of the artifact that is tested simultaneously by different practitioners in different localities (Iivari & Venable, 2009). Thus, for the testing of solutions identified in the form of artifacts, the mental model of ADR adds new possibilities of actions to be considered by the researcher who has as reference the AR approach.

It is important to recognize that the epistemological stance of the researcher in relation to the actors involved is more open and comprehensive in the context of the RA approach than in the DSR approach. In the RA approach, we start from the non-determinism of the actors studied, and multiple aspects of the intervention can be discussed. The training of the researcher who has a broad domain in the AR approach in the assumptions of the DSR, through the ADR, should not restrict this important opening, on the contrary, it should further expand the set of possible scenarios, considering the possibility of generating new knowledge according to the assumptions of design theory. Thus, the researcher as a broad domain of the RA, which turns to the solution of problems of a given locality, may from the foundations of the DSR, work with the possibility of specifying one or more artifacts, with specific purposes and target audience. Thus, the artifacts that are also likely to be generated by RA, can be presented in a more structured and ontologically more effective way in terms of being more easily recognized and understood, thus promoting their reuse by other practitioners, as well as their evolution by other researchers.

GROUNDED DESIGN (GD)

Rohde et al. (2017) proposed the GD method aiming to overcome some difficulties faced by interventionist approaches, such as self-referentiality and contingency. The self-referentiality stance results in a closed system, instead of a system–environment relationship, the organization develops a self-centered system, system-system. The open system allows exchanges with the external environment, facilitates the inclusion of enhancements and advances, while the self-referential system is closed and therefore much more difficult to accept exchanges with the external environment. The artifacts proposed by the researchers through the DSR may be perceived by the practitioners of the organization (the insiders) as an element external to the organization. Hence the recommendations of the ADR in involving as early as possible and with the greatest possible intensity the users/practitioners and other stakeholders of the artifact, with the purpose of having an artifact perceived as being of the organization. The term contingency is related to the issue of social dependence, i.e., the acceptance and use of the artifact goes through the appropriation of the artifact by people of the organization.

The GD approach can be understood as a triangulation between DSR and Grounded Theory (GT) (Corbin & Strauss, 1990; Glaser & Strauss, 1967). Just as Baskerville and Pries-Heje (1999) proposed the addition of GT analysis and coding techniques to the Action Research method, with the aim of having an approach with more rigor and reliability to the theory formulation process, Rohde et al. (2017) proposed the triangulation between DSR and GT to provide a method that allows an insertion of technological artifacts with less rejection and more acceptance by the community of users of the organization. The GD can be understood as a set of principles for the development of artifacts, described in Table 7.1, which are supported by concepts from the practice theory. In the next subsection we will discuss the central concepts of the practice theory.

Table 7.1 Grounded design principles

Principle	Description
Pre-study/context study	Designers must have a strong involvement and insertion with the community of future users, i.e., those who experience the difficulty that is the object of the artifact, in order to understand their social practices
Working on the artifact	The artifact design should be understood as an appropriation process of the artifact by future users. For this, it is essential that the inputs, actions, and outputs of the new artifact are interpreted by users within the context of their social practices. The central idea here is to ensure the use and effectiveness of the artifact
Working with the artifact	To evaluate the usefulness and usability of the new functions of the artifact users must make use of it. From the use, they try to make sense of the new functions to perform their work, considering new ways of performing their jobs, as well as new improvements needed to the artifact under development. It is established this way a process of *learning by designing*
Building the knowledge base	Each tentative design of the artifact should have the results of its use duly observed and recorded in the form of a *design case study* (DCS). This record should contain information on the design options considered, as well as the appropriation process and the effectiveness of the artifacts' functions and the emerging new social practices
Meta-analysis	The results of several DCS registered in the knowledge base can undergo meta-analysis with the aim of identifying cross-sectional similarities and differences. This can lead to some common patterns or design features (structural configurations) that can be typified for the artifact already contemplating specific properties or requirements of social practices for necessary appropriation activities
Evolutionary project organization	Create the organizational culture of developing and updating artifact designs in a strongly participatory manner with users and paying attention to the requirements of social practices for necessary appropriation activities

Source Rohde et al. (2017)

Practice Theory

The artifacts developed from the DSR, ADR or GD can be operated within the context of a personal tool, for example, by a liberal professional, or be directed to a community of professionals who collaborate within an organization or a set of organizations that act in network. The wider the community of people involved for the operationalization of the artifact, the more complex and challenging will be the process of appropriation of the artifact by the community of practitioners. The GD turns to this aspect of the effective use of the artifact by the practitioner(s), which implies most often in behavior change, characterized by the abandonment of one practice and the adoption of another. This transformation of the practice developed by a professional requires not only the replacement of one tool for another, but is characterized as a systemic change that involves social structures and relationships.

Practice theory helps practitioners of pragmatic approaches such as DSR, ADR and GD, to understand the various systemic components involved in the effective appropriation of an artifact by a social group. According to the practice theory a working practice involves several dimensions such as: materiality and embodiment, structure, and cognitive-mental processes. According to Giddens (1984) the structure dimension is characterized by rules (shared knowledge). The materiality and embodiment dimension is composed of artifacts, bodies or natural objects that contribute to the formation of practices. The dimension cognitive-mental processes covers non-material aspects, such as emotion and affectivity linked to the practice, configured as codes that characterize for the practitioner the essence of the practice.

The researcher by following the use of the artifact by the practitioners, in terms of what they are doing and saying in relation to the new artifact, constitutes a way to understand and analyze how to overcome the resistance or difficulties arising from the force of habit linked to the internalized routines. The record of this information at each new adjustment in the design of the artifact, through design case study (see principle "Building the knowledge base" in Table 7.1), as proposed by GD, brings important inputs. The application of coding techniques and content analysis for these inputs, as proposed by the Grounded Theory, allows researchers to analyze the effectiveness of the artifact in terms of its appropriation by practitioners.

REFERENCES

Baskerville, R., & Pries-Heje, J. (1999). Grounded action research: A method for understanding IT in practice. *Accounting, Management & Information Technology, 9*(1), 1–23.

Corbin, J., & Strauss, A. (1990). Grounded theory research: Procedures, canons, and evaluative criteria. *Qualitative Sociology, 13,* 3–21.

Creswell, J. W. (2014). *Research design: Qualitative, quantitative, and mixed methods approaches* (4th ed.). Sage.

Giddens, A. (1984). *The constitution of society.* Polity Press.

Glaser, B. G., & Strauss, A. L. (1967). *The discovery of grounded theory: Strategies for qualitative research.* Hawthorne, NY: Aldine de Gruyter.

Goldkuhl, G. (2013). Action research vs. design research: Using practice research as a lens for comparison and integration. *SIG Prag workshop on IT artefact design & work practice improvement* (5 June 2013), Tilburg, The Netherlands.

Hevner, A. R., March, S. T., Park, J., & Ram, S. (2004). Design science in information systems research. *MIS Quarterly, 28,* 75–105.

Lewin, K. (1946). Action research and minority problems. *Journal of Social Issues, 2*(4), 34–46.

Iivari, J., & Venable, J. R. (2009). Action research and design science research—Seemingly similar but decisively dissimilar. *ECIS 2009 Proceedings* [Online], p. 73. Available at: http://aisel.aisnet.org/ecis2009/73.

Mullarkey, M. T., & Hevner, A. R. (2019). An elaborated action design research process model. *European Journal of Information Systems, 28*(1), 6–20.

Papas, N., O'Keefe, R. M., & Seltsikas, P. (2012). The action research vs design science debate: Reflections from an intervention in eGovernment. *European Journal of Information Systems, 21*(2), 147–159.

Rohde, M., Brödner, P., Stevens, G., Betz, M., & Wulf, V. (2017). Grounded design—A praxeological IS research perspective. *Journal of Information Technology, 32*(2), 163–179.

Romme, A. G. L. (2003). Making a difference: Organization as design. *Organization Science, 14*(5), 558–573.

Sein, M. K., Henfridsson, O., Purao, S., Rossi, M., & Lindgren, R. (2011). Action Design Research. *MIS Quarterly, 35*(1), 37–56.

Susman, G. I., & Evered, R. D. (1978). An assessment of the scientific merits of action research. *Administrative Science Quarterly, 23*(4), 582–603.

Walls, J. G., Widmeyer, G. R., & El Sawy, O. A. (1992). Building an information system design theory for vigilant EIS. *Information Systems Research, 3*(1), 36–59.

CHAPTER 8

Communication of the Results of the DSR Survey

In the fifth chapter, some quite specific aspects for the communication of scientific findings associated with the DSR method were described. To address these specificities Peffers et al. (2007) defined the phase "6. Communication." Among these specific aspects of communication of DSR, two stand out. The first is that DSR should be communicated not only to researchers, but also to practitioners, in order to enhance the return on research (Hevner et al., 2004). The second aspect is regarding the structural specificities of the texts disseminating DSR findings in relation to the structure of texts normally employed for disseminating the findings of traditional positivist or hypothetico-deductive research. DSR adds to the structure of traditional research (Introduction, Literature review, Method, Result, Discussion, and Conclusion (Sun & Linton, 2014)) two other text sections: "artifact description" and "evaluation/demonstration." Obviously the structure of the research project is equally impacted, considering the specificities of the approach and the article. In this chapter we will discuss these various distinctive aspects of communicating DSR, starting with the structural specificities of the text sections of the article communicating DSR findings.

J. O. De Sordi, *Design Science Research Methodology*, https://doi.org/10.1007/978-3-030-82156-2_8

STRUCTURING OF SCIENTIFIC ARTICLES
FOR DISSEMINATION OF DSR FINDINGS

The structure of the article for dissemination of scientific discoveries developed with the DSR approach usually consists of the following chapters: Introduction, Literature review, Method, Artifact description, Evaluation/demonstration, and Discussion. In this subsection the contents addressed in each of these text sections of the DSR article are presented, associating them with the phases of the DSRM approach (studied in the fifth chapter) in which such content is discussed and generated.

Introduction. The introduction chapter should mention some central elements, such as: the community of practitioners, the problem experienced by this community, and the central function of the artifact being proposed. Here the problem should be presented using, preferably, indicators that are known and have sensitive results for the community of practitioners of the artifact. The more widespread and valued the indicator for a given community of professionals, the better the reader's understanding of what is being proposed. If we are proposing an artifact whose primary function is to assist entrepreneurs, a good indicator is the indicator of time/longevity of new companies in a given region. The statement that 50% of new organizations close their activities before completing 24 months is something sensitive to researchers and practitioners associated with entrepreneurship and small business management. If we are proposing a less intrusive surgical equipment, the indicator of length of stay in the intensive care unit in the postoperative period or the indicator of hospital infection levels may be interesting indicators to be used. Thus, we will meet a central concern of every researcher, to answer the question "so what?" that guides the initial work of the editor during the desk review. The introduction needs to highlight to the reader the consequences of the research result as important. In the previous examples, the fact that the artifact helped reduce by 10% the mortality of people or the closure of companies means preserving, respectively, as many lives or as many jobs in a given location in a given period.

Some journals prevent the beginning of the article to have any name for the first section, asking researchers to start writing the first paragraph of the article, right after the presentation of the article title. However, most journals allow the writing of a name for the first section. An interesting aspect is for DSR is to develop a title for the first chapter that is meaningful in relation to its contribution to the practitioner community, something that is more meaningful than just the term "Introduction." Returning to the example of the method-type artifact I presented in the second chapter, the extensive texts cohesion analysis approach or AnaCoTEx (De Sordi et al., 2016), we have that the title of the first chapter made use of this strategy, being defined as: "The Challenge of Extensive and Cohesive Texts."

The more focused and specific the artifact in terms of function performed, the easier it is to identify its target audience, practitioners, as well as possible performance indicators. It is important to remember that in the DSR we do not always have the improvement or continuity of existing artifacts. As observed in the second chapter, there are also actions of invention and exaptation and, in these cases, innovation is even in the environment of application of the artifact. Perhaps the practitioners have not even perceived a problem or a need, i.e., we are talking about an opportunity. For this last case, the publication will only occur if the DSR is well developed, with extensive testing of the artifact in the field, in which it can evidence important advances in important indicators. It would not only be presenting a possible and imaginable advance, but a reality already observed in the field. At this point it is important to remember that editors and reviewers are more sensitive and favorable to publish articles about known problems than solutions to explore opportunities or avoid potential problems.

Thinking about the relationship between the contents of the text sections of the article with the six phases of DSRM (Peffers et al., 2007) presented in the fifth chapter, the texts of this introductory section come mostly from the actions developed in the first two phases of DSRM: phase (1) Problem identification and motivation; and phase (2) Define the objectives for a solution.

Literature review. The literature review section is used in the DSR to describe the various components used for the development of the new artifact. Depending on the creative tactic employed for the proposition of the new artifact (see Fig. 2.2), one may have different levels of ontological challenges in relation to the description of the inputs used to compose the artifact. The creative tactic new combination can aggregate components from different areas of science. Taking as an example the method artifact discussed in the second chapter, the extensive texts cohesion analysis approach or AnaCoTEx (De Sordi et al., 2016), it was composed from the interrelation of two concepts: cross-referencing and matrix of relations of network analysis (NA). These two concepts were described in the second section of the article, called "Concepts Necessary for Understanding the Tool." It is important to highlight the name used for this section in the DSR. One can easily identify a diversity of other names more appropriate to the DSR context than the denomination "literature review" commonly employed in the traditional positivist research structure.

To define how much to describe and conceptualize for each of the constituent components of the artifacts, researchers must reflect on the profile of the potential readership. To this end, it is essential to identify as soon as possible the journal in which the findings of the DSR are intended to be published. From this, ascertain the profile of editors and readers of this journal, to then consider which constituent elements there may be need for further description to the readership. The same reasoning applies to researchers in training, developing their research to obtain the title of master/doctor. In this case, the analyses are made considering the mastery of the concepts by the teacher-researchers who will evaluate their research.

It is important to note that entities to be described and conceptualized in DSR may be of the most diverse. Taking as an example, artifacts proposed by creative action of the exaptation type, evidencing will not be of constituent components but of evidencing the artifact's essential competence or function that is being exapted from one context to another. The description of this scenario of origin, the potentialities of the artifact in its original context, as well as the characterization of this artifact in terms of constituent elements is usually made and presented also in the literature review section. Here we should also think about the readership, how much they should know about the functions of the artifact being exapted, i.e., their level of knowledge regarding the use and application in the context of origin. Regardless of the type of creative action and the type of artifact, the most important thing is for the researcher to be aware of the need to identify and analyze the reader-public profile as soon as possible. Once the artifact and the readership are identified, the researcher must consider the level of mastery of this audience on each of the components necessary for the understanding of the artifact being proposed. Thus, the researcher will know which entities to conceptualize and how much to conceptualize.

Thinking about the relationship between the contents of the text sections of the article with the six phases of DSRM (Peffers et al., 2007) presented in the fifth chapter, the texts of this second section of Literature Review come from the actions developed in the third phase of DSRM:

phase 3, Design and development [of the artifact]. Considering the recursiveness of the DSRM method in terms of field validation, obviously the following Demonstration and Evaluation sections bring the researchers back several times to the Design and development phase [of the artifact]. This is where changes and introductions of components for the best performance of the artifact are discussed.

Method. Considering the few years of existence of the DSR, an aspect to be considered by its practitioners is the mastery of the approach by the editors and reviewers of the journal in which it is intended to publish. Generally, when the research strategy has its first appearance in a given journal it is very common for the authors of the article to have to describe and explain in greater detail the method employed. As an example, Schultze (2000) was the first to publish the findings of a research conducted with the Confessional Ethnography approach in the journal Management Information System Quarterly. Because of this, the author created a predecessor section to Method called "Ethnography and Confessional Writing." Thus, it is important for practitioners of DSR to ascertain the level of mastery of the DSR approach in the journal in which they intend to publish, assessing the need for the development of this additional and predecessor text section to Method. If the need is identified, the texts of the first chapters of this book may be useful for generating introductory content to the DSR approach.

In the method section of the research DSR the most important is to indicate the procedures employed to enable the use of the artifact by practitioners, also highlighting how the results generated from the use of the artifact by practitioners are captured and analyzed. Here several demands come into play, such as: criteria for identification and invitation to practitioners; means employed to enable practitioners to use the artifact; time of use of the artifact by practitioners; place and manner of use of the artifact by practitioners; instruments to collect data associated with the use of the artifact by practitioners; and techniques for analysis of the data associated with the use of the artifact. If only the demonstration of the artifact occurs, the method section must be adapted for this context, thinking about the collection of perceptions of the practitioners who participated in the demonstration. At this point it is important to remember that the DSRM provides for publication of artifacts that have advanced to the demonstration, i.e., not necessarily reached an evaluation phase that would be the ideal situation (Peffers et al., 2007).

As for the tests for artifact evaluation, having as main input the data collected from the use of the artifact by practitioners, it is never too much to emphasize the necessary care. In this sense, Hevner et al. (2004, p. 83) when defining seven guidelines for conducting the DSR, reserved two of them specifically for the artifact evaluation:

> Design evaluation −The utility, quality, and efficacy of a design artifact must be rigorously demonstrated via well-executed evaluation methods; Research rigor −Design-science research relies upon the application of rigorous methods in both the construction and evaluation of the design artifact.

As for the association of contents of the text sections of the article with the six phases of DSRM (Peffers et al., 2007), presented in the fifth chapter, we have that the texts of this Method section come mostly from two phases of DSRM: phase 4, Demonstration; and/or phase 5, Evaluation.

Artifact description. Walls et al. (1992) when describing Design Theory to his peers in the area of information systems highlighted two components necessary for the development of knowledge according to DSR, the "design product" and the "design process." The design product, encompassing four contents, the meta-requirements, the meta-design, the kernel theories (governing design requirements), and the testable design product hypotheses; and the design process encompassing three contents, the design method, the kernel theories (governing design process itself), and the testable design process hypotheses. Subsequently, Gregor and Hevner (2013) called the term design product synonymous with "design artifact" and the term design process synonymous with "design search (development)."

The descriptive information of the design product should allow the typical reader of the journal in which the article is published to understand, reproduce, and evolve the artifact (redesign) if necessary. For this there should be great clarity in the description of the meta-requirements and meta-designs of the artifact. These descriptions may take different forms depending on the many subtypes available for each of the artifact types: construct, method, model, and instantiation. In Table 5.1, we present the meta-requirements and meta-designs for an artifact of the construct type, more specifically of the taxonomy subtype. The third column of Table 5.1, called "Contextualization (in the design process that was achieved)," was inserted more for didactic purposes, to give more understanding to the reader, and may be dispensed with in the context of a scientific article. Here, due to the great diversity of types and subtypes of artifacts, there are no templates or standards for the description of meta-requirements and meta-designs. It is observed at this point the ontological sensitivity necessary to researchers, editors and reviewers, regarding the level of specification required to the meta-requirements and meta-designs in order to allow the reconstruction, use and evolution of the artifact by other researchers. As for the description of the design process (or design search (development)) it is often omitted from the scientific article due to space limitation in terms of the number of words. It ends up configuring itself more as a history of the creative process employed by researchers, very important for the realization of the research, but not as something essential for its continuity.

Thinking about the relationship between the contents of the text sections of the DSR article structure with the six phases of DSRM (Peffers et al., 2007) presented in the fifth chapter, the contents of this fourth section of the DSR article, the "artifact description," come from actions developed in the third phase of DSRM: phase 3, Design and development [of the artifact]. It is important to note that the texts in the Literature review section of the article also come from this same phase 3 of DSRM. This makes perfect sense, considering that the, "kernel theories (governing design requirements)" is one of the four components of the design product (Walls et al., 1992). Although the four components of the design product (meta-requirements, meta-design, kernel theories, and testable design product hypotheses) are thought and processed in a simultaneous and integrated way, for purposes of presenting the findings of the DSR in the form of article, they are presented in different sections and at different times.

Evaluation/demonstration. The great attractions of the DSRM are realized by the use of the artifact in the field by the practitioners, i.e., the evaluation is the priority, being the demonstration a second option when the first one is impossible. The conduction of the field evaluation, in the real context, should highlight the observation and the respect for four central aspects as discussed in the DSRM Evaluation phase: (a) use of the artifact, (b) tester user, (c) environment of use, and d) inputs for use (see Table 5.2 for description of each of these aspects). In the Method section it was described the operational procedures for selecting practitioners, for collecting data from the use of the artifact by practitioners, among several other procedures. In this section the results are presented, the information linked to each of the four central aspects of evaluation, i.e., answering the guiding questions of the four central aspects described in Table 5.2. This involves presenting information about: (a) names and/or descriptive of practitioners who made use of the artifact; (b) information of training and other actions of preparation of practitioners for the use of the artifact; (c) information of the availability of the artifact and the period of use by practitioners; (d) information of the place and form of use by practitioners; (e) data and information captured from the use of the artifact, contextualizing through indicators that allow comparison with the existing artifacts.

The analysis of the results may involve since descriptive statistics for comparison of indicators that allow comparing the proposed artifact to the historical results of existing artifacts, as well as the application of various statistical techniques for analysis of the data collected in the field. The comparative axis is more effective when it is possible to work with important indicators and widely disseminated and known by practitioners, this makes the communication of the evaluation process more simple and direct. Unfortunately, these indicators are not always available for the core function of the artifact. This was evident in section two, when discussing the "usefulness of the artifact," when we explored the indicator adopted for analysis of the artifact we are describing as an example, the extensive texts cohesion analysis approach or AnaCoTEx (De Sordi et al., 2016). Although the problem was experienced by many practitioners (consultants, auditors, editors, and researchers), there was no indicator in use among practitioners for analyzing cohesion between parts of an extensive text. The solution was to adopt an indicator linked to the kernel theories employed for the development of the artifact, more specifically the variable "network density" of network analysis (Wasserman & Faust, 1994).

The more complex the artifact, the greater the quantity of indicators and data to be collected so as to holistically analyze the various multifaceted aspects: financial, environmental, social, operational, and qualitative dimensions, among others. Another difficulty in the performance of artifacts evaluation occurs when they propose to meet rare and time-consuming operations. In these situations, many times the evaluation ends up being the analysis of a single case, covering the processing of only one instance. For these situations of greater operational difficulty for execution of the evaluation procedures, it can be considered alternatively the demonstration of the artifact. Another aspect that may lead to the demonstration instead of evaluation is a very innovative artifact, which may demand many discussions on the social side in terms of acceptance and appropriation of it by the practitioners. Thus, as Gregor and Hevner (2013, p. 351) pointed out, in extreme situations in terms of innovation, involving "very novel artifacts, a 'proof-of-concept' may be sufficient."

As for the association of contents of the text sections of the article with the six phases of DSRM (Peffers et al., 2007), presented in the fifth chapter, we have that the texts of this Method section come from one of two phases of DSRM: phase 4, Demonstration; or phase 5, Evaluation.

Discussion. In the discussion section the researchers and artifact proposers should present and discuss the results achieved by practitioners from the use of the artifact. One can think about the positive and negative aspects or, in another perspective, the impacts provided to the beneficiaries from the function performed by the artifact, as well as the efforts required for the use of the artifact by the practitioners. Whenever possible these results should be compared with the results of other artifacts in use for the exercise of the same function. This section should demonstrate the efforts of the researchers to collect and compare the results for each of the important dimensions, taking into consideration the core function of the artifact.

As discussed in phase 3 of DSRM, "Design and development," the artifact design is being defined and improved through an interactive process, called by Hevner et al. (2004) as generate/test cycle. In this sense, partial results of the indicators, for different evolutionary versions of the artifact, can be useful to characterize some important specificities incorporated to the artifact design. If the artifact requires human intervention, which is very common, issues associated with practitioners' perception is fundamental. As discussed in the seventh chapter, in the description of Grounded Design, we must bring into the discussions issues associated with the appropriation and use of the artifact by the community of practitioners, i.e., a vision not only technological, but techno-social. The premise is that it is not enough just to develop a good artifact in terms of efficiency, the application and use of this artifact by practitioners is what will actually provide gains to society.

At the end of the discussion section, in a conclusion subsection, it is important to return to the type or class of problem defined in the Introduction. The artifact should be positioned as an instance of the possible solutions, explaining the superior performance to the others by the synthesis of its distinctive aspects in terms of design. Limitations of the artifact (design product) should be described and pointed out as opportunities for further research. Limitations of the design process should be reported if they are associated with limitations of the artifact, in order to assist in the review of the method by those who may continue the research.

DSR Project Structuring

In addition to the scientific article, another document of a scientific nature written by researchers is the research project. As observed for the scientific article, the structure of the research project that adopts the DSR approach also has its specificities. Nagle and Sammon (2016) proposed a template for quality validation of DSR projects, called by them "The Design Research Canvas." The template is structured based on four themes associated with the phases or work fronts of DSR: (a) problem, (b) design and build, (c) evaluation, and (d) contributions. For each of these four themes, two guiding questions are presented, one for the researchers' context and the other for the practitioners' context. Thus, we have 8 guiding questions, which are described in Table 8.1. From this analytical structure proposed by Nagle and Sammon (2016), it is possible to validate not only the relevance of the DSR project to the theme to be researched, but also the text structure in terms of contents necessary to write the research project.

Table 8.1 Questions for analyzing DSR projects

Perspective	Practioner	Researcher
Problem	Worth solving?	Worth researching?
Design and build	Well organised?	Well documented?
Evaluation	Role of artefact in problem solution?	Reflective learnings from the study?
Contributions	So what (for practioners)?	So what (for researchers)?

Source Adapted from Nagle and Sammon (2016)

Dissemination of the Artifact to Practitioners

Few research strategies make space for direct communication with practitioners, non-academics, as with DSR. Because of this, researchers are trained and accustomed to disseminating their findings only to their peers, to other researchers in the scientific community. In this context, of a researcher preparing text for other researchers, although there is an ontological challenge of announcing the new, it is minimized because the researcher is writing for a known audience, of which he is part. It is important to remember that researchers, before developing their scientific texts, have practiced a lot of reading of scientific texts that is, they have an experience of readers of what they should produce.

Institutions, research centers and researchers that adopt the DSR strategy need to prepare for the elaboration of another type of communication, the non-scientific, that directed to the practitioners for which the artifact is intended. The practitioner audience has a more functionalist view, directed at how to use the artifact. As discussed in the first chapter, in the subsection "Literature for practitioners," the text for practitioners is quite different from the communication developed for researchers in the form of scientific article. Communication for practitioners should cover topics such as: a) difficulty addressed by the artifact, indicating the problem of knowledge of the practitioner community in order to motivate the reading of the text; b) present the artifact, highlighting its innovative features and the differential in terms of results; c) describe how to apply or use the artifact. Here there is no need to describe data collection and analysis methods. The main criteria of practitioners for choosing texts for reading are: pertinence with the need and the credibility of the channel, that is, the perception they have of that communication channel.

Institutes and research centers that want to extract the full potential of research developed with the DSR should have communications teams helping researchers to communicate with practitioners. Here is the same challenge already well described by academia with writing demands for the patenting process of scientific discoveries. Researchers hardly have the skills to conduct these two types of writing. Generally, researchers rely on support teams for writing and administrative and technical routing of patent applications. In the case of the communication of the results of the DSR to the public of practitioners, we observe the same situation, a very technical activity and distinct from the skills of researchers, and it would be more appropriate to have professionals specialized in this type of non-scientific communication.

REFERENCES

De Sordi, J. O., Meireles, M., & de Oliveira, O. L. (2016). The Text Matrix as a tool to increase the cohesion of extensive texts. *Journal of the Association for Information Science and Technology, 67*(4), 900–914.

Gregor, S., & Hevner, A. (2013). Positioning and presenting design science research for maximum impact. *MIS Quarterly, 37*(2), 337–355.

Hevner, A. R., March, S. T., Park, J., & Ram, S. (2004). Design science in information systems research. *MIS Quarterly, 28*, 75–105.

Nagle, T., & Sammon, D. (2016). The development of a Design research canvas for data practitioners. *Journal of Decision Systems, 25*, 369–380.

Peffers, K., Tuunanen, T., Rothenberger, M. A., & Chatterjee, S. (2007). A design science research methodology for information systems research. *Journal of Management Information Systems, 24*(3), 45–77.

Schultze, U. (2000). A confessional account of an ethnography about knowledge work. *MIS Quarterly, 24*(1), 3–79.

Sun, H., & Linton, J. D. (2014). Structuring papers for success: Making your paper more like a high impact publication than a desk reject. *Technovation, 34*(10), 571–573.

Walls, J. G., Widmeyer, G. R., & El Sawy, O. A. (1992). Building an information system design theory for vigilant EIS. *Information Systems Research, 3*(1), 36–59.

Wasserman, S., & Faust, K. (1994). *Social network analysis: Methods and applications*. Cambridge University Press.

Annex A: Basic Heuristics

The set of basic heuristics for the problem solving according to Savransky (2000, p. 13):

> NEOLOGY (originating from the Latin for 'novelty' or 'knowledge of new') consists of the new application of established processes, construction, shape, material, and their properties, etc. This process, etc. is not new but its application to the specific field is;
>
> ADAPTATION includes fitting of known processes, constructions, shapes, materials, and their properties to specific conditions of labour;
>
> MULTIPLICATION of system functions and parts and multiplied systems remain similar to each other, of the same type. Multiplication includes not only methods associated with enlargement of characteristics (hyperbolisation) but also with their miniaturisation;
>
> DIFFERENTIATION of system functions and elements: functional links between system elements weaken; elements of construction and working processes become spatially and temporarily separated;
>
> INTEGRATION includes joining, combining, reducing in number, and simplifying of functions and forms of elements and the system as a whole: production and construction elements and working processes become spatially and temporarily closer;
>
> INVERSION is a reversion of functions, shapes, and mutual position of elements and the system as a whole;

J. O. De Sordi, *Design Science Research Methodology*,
https://doi.org/10.1007/978-3-030-82156-2

PULSATION encompasses the group of design and creative methods associated with changes in process continuity. Pulses may repeat periodically or aperiodically; a pulse may also be single (method of overshoot);

DYNAMISATION suggests which element parameters or techniques as a whole must be changeable and optimal at every stage of a process or in a new mode.

ANALOGY is using similarity or resemblance of some aspects of systems (objects, phenomena) which are otherwise different as a whole;

IDEALISATION is presenting an ideal solution as an aim to be achieved to make a start from the best result.

REFERENCES

Adam, M. T. P., Krämer, J., Jähnig, C., Seifert, S., & Weinhardt, C. (2011). Understanding auction fever: A framework for emotional bidding. *Electronic Markets, 21*(3), 197–207.

Afflerbach, P., Bolsinger, M., & Röglinger, M. (2016). An economic decision model for determining the appropriate level of business process standardization. *Business Research, 9*(2), 335–375.

Ahuja, G., & Lampert, C. M. (2001). Entrepreneurship in the large corporation: A longitudinal study of how established firms create breakthrough inventions. *Strategic Management Journal, 22,* 521–543.

Aloysius, J. A., & Venkatesh, V. (2013). *Mobile point-of-sale and loss prevention: An assessment of risk.* Retail Industry Leaders Association [WWW document]. http://waltoncollege.uark.edu/lab/JAloysius/RILA%20Report/MobilePOSReport.pdf.

Andriani, P., & Carignani, G. (2014). Modular exaptation: A missing link in the synthesis of artificial form. *Research Policy, 43*(9), 1608–1620.

Astor, P. J., Adam, M. T. P., Jercic, P., Schaaff, K., & Weinhardt, C. (2013). Integrating biosignals into information systems: A NeuroIS tool for improving emotion regulation. *Journal of Management Information Systems, 30*(3), 247–277.

Asselin, M. E. (2003). Insider research: Issues to consider when doing qualitative research in your own setting. *Journal for Nurses in Staff Development, 19*(2), 99–103.

Baskerville, R. L., Kaul, M., & Storey, V. C. (2015). Genres of inquiry in design science research: Justification and evaluation of knowledge production. *MIS Quarterly, 39*(3), 541-A9.

© The Author(s) 2021
J. O. De Sordi, *Design Science Research Methodology*,
https://doi.org/10.1007/978-3-030-82156-2

Baskerville, R., & Pries-Heje, J. (1999). Grounded action research: A method for understanding IT in practice. *Accounting, Management & Information Technology, 9*(1), 1–23.

Bemelmans, J., Voordijk, H., & Vos, B. (2013). Designing a tool for an effective assessment of purchasing maturity in construction. *Benchmarking, 20*(3), 342–361.

Bittner, E. A. C., & Leimeister, J. M. (2014). Creating shared understanding in heterogeneous work groups: Why it matters and how to achieve it. *Journal of Management Information Systems, 31*(1), 111–144.

Bjarnason, E., Sharp, H., & Regnell, B. (2019). Improving requirements-test alignment by prescribing practices that mitigate communication gaps. *Empirical Software Engineering, 24*, 2364–2409.

Brady, D. A., Tzortzopoulos, P., Rooke, J., Carlos, T. F., & Tezel, A. (2018). Improving transparency in construction management: A visual planning and control model. *Engineering, Construction and Architectural Management, 25*(10), 1277–1297.

Bunge, M. A. (1974). *Treatise on Basic Philosophy Volume 2: Semantics II—Interpretation and truth.* Dordrecht, The Netherlands: Kluwer Academic Publishers.

Burgoyne, J., & James, K. T. (2006). Towards best or better practice in corporate leadership development: Operational issues in mode 2 and design science research. *British Journal of Management, 17*(4), 303–316.

Collins. (2013). *English dictionary.* http://www.collinsdictionary.com/dictionary/english/cross-reference.

Comesaña-Campos, A., Casal-Guisande, M., Cerqueiro-Pequeño, J., & Bouza-Rodríguez, J.-B. (2020). A methodology based on expert systems for the early detection and prevention of hypoxemic clinical cases. *International Journal of Environmental Research and Public Health, 17*(22), 8644.

Corbin, J., & Strauss, A. (1990). Grounded theory research: Procedures, canons, and evaluative criteria. *Qualitative Sociology, 13*, 3–21.

Creswell, J. W. (2014). *Research design: Qualitative, quantitative, and mixed methods approaches* (4th ed.). Sage.

Danneels, E. (2002). The dynamics of product innovation and firm competencies. *Strategic Management Journal, 23*(12), 1095–1121.

David, P. A., Mowery, D., & Steinmueller, W. E. (1992). Analysing the economic payoffs from basic research. *Economics of Innovation and New Technology, 2*(1), 73–90.

De Sordi, J. O., Meireles, M., & de Oliveira, O. L. (2016). The Text Matrix as a tool to increase the cohesion of extensive texts. *Journal of the Association for Information Science and Technology, 67*(4), 900–914.

De Sordi, J. O., Nelson, R. E., Meireles, M., Hashimoto, M., & Junior, M. D. F. C. (2020). A longitudinal study of the creation methods used by entrepreneurs to develop new products and services. *International Journal of Entrepreneurship and Innovation Management, 24*(6), 482–502.

De Sordi, J. O., Nelson, R. E., Meireles, M., Hashimoto, M., & Rigato, C. (2019). Exaptation in management: Beyond technological innovations. *European Business Review, 31*, 64–91.

Dew, N., & Sarasvathy, S. D. (2016). Exaptation and niche construction: Behavioral insights for an evolutionary theory. *Industrial and Corporate Change, 25*(1), 167–179.

Dijkman, R., Rosa, M. L., & Reijers, H. A. (2012). Managing large collections of business process models-current techniques and challenges. *Computers in Industry, 63*(2), 91–97.

Dos Santos, J. C., & Da Silva, M. M. (2015). Price management in IT outsourcing contracts. The path to flexibility. *Journal of Revenue and Pricing Management, 14*(5), 342–364.

Duggan, G. B., & Payne, S. J. (2009). Text skimming: The process and effectiveness of foraging through text under time pressure. *Journal of Experimental Psychology Applied, 15*(3), 228–42.

Duke University. (2013). *Graduate school, scientific writing resources*. https://cgi.duke.edu/web/sciwriting/index.php?action=lesson2#principles

Dwyer, S. C., & Buckle, J. L. (2009). The space between: On being an insider-outsider in qualitative research. *International Journal of Qualitative Methods, 8*(1), 54–63.

Ebel, B., & Hofer, M. B. (2016). *Automotive management—Strategie und marketing in der automobilwirtschaft* (2nd Ed.). Berlin/Heidelberg, Germany: Springer.

Eco, U. (1989). *Come si Fa una Tesi di Laurea*. Milano, Italy: Bompiani.

Ellis, T. J., & Levy, Y. (2008). Framework of problem-based research: A guide for novice researchers on the development of a research-worthy problem. *Informing Science, 11*, 17–33.

Font, J. M., Hedvall, A., & Svensson, E. (2017). Towards teaching maternal healthcare and nutrition in rural Ethiopia through a serious game. In *Extended abstracts publication of the annual symposium on computer-human interaction in play* (pp. 187–193): Association for Computing Machinery.

Garud, R., Gehman, J., & Giuliani, A. P. (2016) Technological exaptation: A narrative approach. *Industrial and Corprate Change, 25*, 149–166.

Genemo, H., Miah, S. J., & McAndrew, A. (2016). A design science research methodology for developing a computer-aided assessment approach using method marking concept. *Education and Information Technologies, 21*(6), 1769–1784.

Giddens, A. (1984). *The constitution of society*. Polity Press.

Giménez, Z., Mourgues, C., Alarcón, L. F., Mesa, H., & Pellicer, E. (2020). Value analysis model to support the building design process. *Sustainability, 12*(10), 4224.

Glaser, B. G., & Strauss, A. L. (1967). *The discovery of grounded theory: Strategies for qualitative research.* Hawthorne, NY: Aldine de Gruyter.

Goes, P. B. (2014). Design science research in top information systems journals. *MIS Quarterly, 38*, iii–viii.

Goldkuhl, G. (2012). Design research in search for a paradigm: Pragmatism is the answer. In M. Helfert & B. Donnellan (Eds.), *Practical aspects of design science* (Vol. 286, pp. 84–95). Springer.

Goldkuhl, G. (2013). Action research vs. design research: Using practice research as a lens for comparison and integration. *SIG Prag workshop on IT artefact design & work practice improvement* (5 June 2013), Tilburg, The Netherlands.

Gregor, S. (2006). The nature of theory in information systems. *MIS Quarterly, 30*(3), 611–642.

Gregor, S., & Hevner, A. (2013). Positioning and presenting design science research for maximum impact. *MIS Quarterly, 37*(2), 337–355.

Gregor, S., & Hevner, A. (2014). The Knowledge Innovation Matrix (KIM): A clarifying lens for innovation. *Informing Science: THe International Journal of an Emerging Transdiscipline, 17*, 217–239.

Guba, E. G. (1981). Criteria for assessing the trustworthiness of naturalistic inquiries. *Educational Communication and Technology, 29*(2), 75–91.

Haeussler, C., & Assmus, A. (2021). Bridging the gap between invention and innovation: Increasing success rates in publicly and industry-funded clinical trials. *Research Policy, 50*(2), 104155.

Hatchuel, A., & Weil, B. (2009). C-K design theory: Na advanced formulation. *Research in Engineering Design, 19*, 181–192.

Hedvall, A., & Svensson, E. (2017). *Teaching maternal healthcare and nutrition in rural Ethiopia through a serious game* (Dissertation, Malmö högskola/Teknik och samhälle). http://urn.kb.se/resolve?urn=urn:nbn:se:mau:diva-20939.

Hevner, A. R., March, S. T., Park, J., & Ram, S. (2004). Design science in information systems research. *MIS Quarterly, 28*, 75–105.

Hoehle, H., Aloysius, J. A., Goodarzi, S., & Venkatesh, V. (2019). A nomological network of customers' privacy perceptions: Linking artifact design to shopping efficiency. *European Journal of Information Systems, 28*(1), 91–113.

Hoehle, H., & Venkatesh, V. (2015). Mobile application usability: Conceptualization and instrument development. *MIS Quarterly, 39*(2), 435–472.

Hoehle, H., Zhang, X., & Venkatesh, V. (2015). An espoused cultural perspective to understand continued intention to use mobile applications: A four-country study of mobile social media application usability. *European Journal of Information Systems, 24*(3), 337–359.

Hoffmann, A., Hoffmann, H., & Söllner, M. (2013). Fostering initial trust in applications—Developing and evaluating requirement patterns for application websites. In *21st European Conference on Information Systems* (ECIS), Utrecht, The Netherlands.

Holler, M., Uebernickel, F., & Brenner, W. (2017, June 5–10). Defining archetypes of e-collaboration for product development in the automotive industry. *Paper presented at the 25th European Conference on Information Systems* (ECIS), Guimarães, Portugal. https://aisel.aisnet.org/ecis2017_rp/8.

Jurczak, J. (2008). Intellectual capital measurement methods. *Economics and Organization of Enterprise, 1*(1), 37–45.

Kanuha, V. K. (2000). "Being" native versus "going native": Conducting social work research as an insider. *Social Work, 45*(5), 439–447.

Khazanchi, D., Murphy, J. D., & Petter, S. C. (2008, May 23–24). *Guidelines for evaluating patterns in the IS domain*. Paper presented at the Third Midwest United States Association for Information Systems (MWAIS), Eau Claire, Wisconsin. https://digitalcommons.unomaha.edu/isqafacproc/7.

Kogut, B., & Zander, U. (1997). Knowledge of the firm: Combinative capabilities, and the replication of technology. In L. Prusak (ed.), *Knowledge in organizations* (pp. 17–35). Boston: Butterworth-Heinemann.

Ku, G., Malhotra, D., & Murnighan, J. K. (2005). Towards a competitive arousal model of decision-making: A study of auction fever in live and Internet auctions. *Organizational Behavior and Human Decision Processes, 96*(2), 89–103.

Kuechler, B., & Vaishnavi, V. (2008). On theory development in design science research: Anatomy of a research project. *European Journal of Information Systems, 17*(5), 489–504.

Laato, S., Laine, T. H., Seo, J., Ko, W., & Sutinen, E. (2017). Designing a game for learning math by composing: A finnish primary school case. In *2017 IEEE 17th International Conference on Advanced Learning Technologies (ICALT)*, Timisoara, Romania, pp. 136–138.

Lapão, L. V., da Silva, M. M., & Gregório, J. (2017). Implementing an online pharmaceutical service using design science research. *BMC Medical Informatics and Decision Making, 17*(1), 1–14.

Lee, J. S., Pries-Heje, J., & Baskerville, R. (2011) *Theorizing in design science research*. Paper presented at the Proceedings of the 6th international conference on Service-oriented perspectives in design science research, Milwaukee, WI, USA.

Lee, C. J., Sugimoto, C. R., Zhang, G., & Cronin, B. (2013). Bias in peer review. *Journal of the American Society for Information Science & Technology, 64*(1), 2–17.

Lewin, K. (1946). Action research and minority problems. *Journal of Social Issues*, 2(4), 34–46.

Iivari, J., & Venable, J. R. (2009). Action research and design science research—Seemingly similar but decisively dissimilar. *ECIS 2009 Proceedings* [Online], p. 73. Available at: http://aisel.aisnet.org/ecis2009/73.

March, S. T., & Smith, G. F. (1995). Design and natural science research on information technology. *Decision Support Systems*, 15(4), 251–266.

McKay, J., Marshall, P., & Hirschheim, R. (2012). The design construct in information systems design science. *Journal of Information Technology*, 27(2), 125–139.

Merton, R. K. (1949). *Social theory and social structure: Toward the codification of theory and research*. Free Press.

Miles, M. B., & Huberman, A. M. (1994). *Qualitative data analysis: An expanded sourcebook* (2nd ed.). Sage.

Mullarkey, M. T., & Hevner, A. R. (2019). An elaborated action design research process model. *European Journal of Information Systems*, 28(1), 6–20.

Nagle, T., & Sammon, D. (2016). The development of a Design Research Canvas for data practitioners. *Journal of Decision Systems*, 25, 369–380.

Negrut, V. (2011). The Europeanization of Public Administration through the general principles of good administration. *Acta Universitatis Danubius. Juridica*, 7(2), 1–15.

Newell, A., & Simon, H. (1972). *Human problem solving*. Prentice Hall.

Ofek, E., & Turut, O. (2008). To innovate or imitate? Entry strategy and the role of market research. *Journal of Marketing Research*, 45(5), 575–592.

Österle, H., & Otto, B. (2009). *A method for consortial research. Institute of Information Management*. University of St. Gallen, Arbeitsbericht Nr. BE HSG/CC CDQ/6.

Österle, H., & Otto, B. (2010). A method for researcher-practitioner collaboration in design-oriented IS research. *Business & Information Systems Engineering*, 2(5), 283–293.

Otto, B., & Österle, H. (2010a). Practical relevance through consortium research? Findings from an expert interview study. In *5th International Conference on Design Science Research in Information Systems and Technology (DESRIST 2010)*, St. Gallen, Switzerland.

Otto, B., & Österle, H. (2010b). Relevance through consortium research? A case study. In *18th European Conference on Information Systems (ECIS 2010)*, Pretoria.

Papas, N., O'Keefe, R. M., & Seltsikas, P. (2012). The action research vs design science debate: Reflections from an intervention in eGovernment. *European Journal of Information Systems*, 21(2), 147–159.

Patton, M. Q. (1987). *How to use qualitative methods in evaluation*. Sage.

Peffers, K., Tuunanen, T., Rothenberger, M. A., & Chatterjee, S. (2007). A design science research methodology for information systems research. *Journal of Management Information Systems, 24*(3), 45–77.

Peffers, K., Tuunanen, T., & Niehaves, B. (2018). Design science research genres: Introduction to the special issue on exemplars and criteria for applicable design science research. *European Journal of Information Systems, 27*(2), 129–139.

Petter, S., Khazanchi, D., & Murphy, J. D. (2010). A design science based evaluation framework for patterns. *Data Base for Advances in Information Systems, 41*(3), 9–26.

Piirainen, K. A., Kolfschoten, G. L., & Lukosch, S. (2012). The joint struggle of complex engineering: A study of the challenges of collaborative design. *International Journal of Information Technology & Decision Making, 11*(6), 1087–1125.

Pries-Heje, J., & Baskerville, R. (2008). The design theory nexus. *MIS Quarterly, 32*(4), 731–755.

Rick, S., & Loewenstein, G. (2008). The role of emotion in economic behavior. In M. Lewis, J. M. Haviland-Jones, & L. F. Barrett (Eds.), *Handbook of emotions* (3rd ed., pp. 138–156). New York: The Guilford Press.

Robinson, O. C. (2011). The idiographic/nomothetic dichotomy: Tracing historical origins of contemporary confusions. *History & Philosophy of Psychology, 13*(2), 32–39.

Rohde, M., Brödner, P., Stevens, G., Betz, M., & Wulf, V. (2017). Grounded design—A praxeological IS research perspective. *Journal of Information Technology, 32*(2), 163–179.

Romme, A. G. L. (2003). Making a difference: Organization as design. *Organization Science, 14*(5), 558–573.

Sampa, M. B., Hoque, M. R., Islam, R., Nishikitani, M., Nakashima, N., Yokota, F., Kikuchi, K., Rahman, M. M., Shah, F., & Ahmed, A. (2020). Redesigning portable health clinic platform as a remote healthcare system to tackle COVID-19 pandemic situation in unreached communities. *International Journal of Environmental Research and Public Health, 17*(13), 4709.

Savransky, S. (2000). *Engineering of creativity: Introduction to TRIZ methodology of inventive problem solving.* Boca Raton, FL: CRC Press.

Schultze, U. (2000). A confessional account of an ethnography about knowledge work. *MIS Quarterly, 24*(1), 3–79.

Sein, M. K., Henfridsson, O., Purao, S., Rossi, M., & Lindgren, R. (2011). Action Design Research. *MIS Quarterly, 35*(1), 37–56.

Shumacher, P. (2012). *The autopoiesis of architecture: A new agenda for architecture* (Vol. 2). Wiley.

Simon, H. A. (1976). The business school: A problem in organizational design. In H. A. Simon (Ed.), *Administrative behavior: A study of decision-making processes in administrative organization* (pp. 335–356). New York: Free Press.

Simon, H. A. (1996). *The sciences of the artificial* (3rd ed.). MIT Press.

Sun, H., & Linton, J. D. (2014). Structuring papers for success: Making your paper more like a high impact publication than a desk reject. *Technovation, 34*(10), 571–573.

Sun, Y., & Kantor, P. B. (2006). Cross-evaluation: A new model for information system evaluation. *Journal of the American Society for Information Science and Technology, 57*(5), 614–628.

Susman, G. I., & Evered, R. D. (1978). An assessment of the scientific merits of action research. *Administrative Science Quarterly, 23*(4), 582–603.

Taura, T., & Nagai, Y. (2012). *Concept generation for design creativity: A systematized theory and methodology*. London: Springer.

Teubner, T., Adam, M., & Riordan, R. (2015). The impact of computerized agents on immediate emotions, overall arousal and bidding behavior in electronic auctions. *Journal of the Association for Information Systems, 16*(10), 838–879.

Tidd, J., Bessant, J., & Pavitt, K. (2005). *Managing innovation: Integrating technological, market and organizational change* (3rd ed.). Chichester, UK: John Wiley & Sons.

Tofan, A. D. (2006). *Instituții administrative europene/European administrative institutions*. Bucharest (RO): C.H. Beck.

Turner, J. R., & Cochrane, R. A. (1993). Goals-and-methods matrix: Coping with projects with ill defined goals and/or methods of achieving them. *International Journal of Project Management, 11*(2), 93–102.

Usener, C. A., Majchrzak, T. A., & Kuchen, H. (2012). E-assessment and software testing. *Interactive Technology and Smart Education, 9*(1), 46–56.

van De Ven, A. H. (2007). *Engaged scholarship: A guide for organizational and social research*. Oxford University Press.

van Aken, J. E. (2005). Management research as a design science: Articulating the research products of mode 2 knowledge production in management. *British Journal of Management, 16*, 19–36.

van Aken, J. E., & Romme, G. (2009). Reinventing the future: Adding design science to the repertoire of organization and management studies. *Organization Management Journal, 6*(1), 2–12.

Venable, J., Pries-heje, J., & Baskerville, R. (2016). FEDS: A framework for evaluation in design science research. *European Journal of Information Systems, 25*(1), 77–89.

Venkatesh, V., Aloysius, J. A., Hoehle, H., & Burton, S. (2017). Design and evaluation of auto-id enabled shopping assistance artifacts in customers' mobile phones: Two retail store laboratory experiments. *MIS Quarterly, 41*(1), 83–113.

Venkatesh, V., Morris, M. G., Davis, G. B., & Davis, F. D. (2003). User acceptance of information technology: Toward a unified view. *MIS Quarterly, 27*(3), 425–478.

Venkatesh, V., Thong, J. Y. L., & Xu, X. (2012). Consumer acceptance and use of information technology: Extending the unified theory of acceptance and use of technology. *MIS Quarterly, 36*(1), 157–178.

Walls, J. G., Widmeyer, G. R., & El Sawy, O. A. (1992). Building an information system design theory for vigilant EIS. *Information Systems Research, 3*(1), 36–59.

Wasserman, S., & Faust, K. (1994). *Social network analysis: Methods and applications*. Cambridge University Press.

Weeding, S., & Dawson, L. (2012). Laptops on trolleys: Lessons from a mobile-wireless hospital ward. *Journal of Medical Systems, 36*(6), 3933–3943.

Yin, R. K. (2014). *Case study research: Design and methods* (5th ed.). Sage.

INDEX

A

abduction, 9, 22

action-reflection cycle, 112–114, 116

administration, 2, 39, 45, 51, 53, 55, 62, 74

artifact description, 24, 71, 74, 121, 122, 127

assumption, 10, 30, 72, 81, 111, 113, 115, 116

C

class of problem, 14, 15, 24, 63, 68, 113–115, 129

co-creation, 115

communication, 31, 34, 59, 61, 62, 64, 68, 69, 73–76, 94, 97, 104, 121, 128, 131

construct, 17, 29–33, 35, 36, 38, 40–42, 64, 74, 81, 86, 91, 97, 98, 100–103, 115, 126

contribution type, 79

creative tactic, 22–24, 39, 65–67, 69, 123

D

demonstration, 17, 36, 37, 59, 70, 71, 73–76, 85, 89, 91, 96, 98, 99, 105, 121, 122, 125–127, 129

design requirement, 75, 76, 126, 127

design theory, 9, 10, 14, 64, 79, 81, 97, 114–116, 126

E

education, 51, 53, 54

engineering, 1, 2, 8, 32, 34, 51, 53, 56, 69, 114

entry point, 37, 41, 42, 106

episode, 81, 96, 98, 101–104

evaluation, 36–39, 42, 54, 55, 59, 64, 70–76, 81, 82, 85, 87, 88, 91, 95, 98, 99, 103, 105, 106, 113, 116, 121, 122, 125–130

evolutionary cycle, 39, 56

exaptation, 19–22, 24, 30, 53, 59, 65, 123, 124

F

functional saturation, 68

The manufacturer's authorised representative in the EU is Springer Nature Customer Service Centre GmbH, Europaplatz 3, 69115 Heidelberg, Germany. If you have any concerns regarding our products, please contact ProductSafety@springernature.com

Printed and bound by CPI Group (UK) Ltd, Croydon, CR0 4YY

24/04/2026

02096343-0001